COURTYARD HOUSING
CULTURAL SUSTAIN

T0227935

Courtyard Housing and Cultural Sustainability

Theory, Practice, and Product

DONIA ZHANG

Oxford Brookes University, UK

Routledge

Taylor & Francis Group

LONDON AND NEW YORK

First published 2013 by Ashgate Publishing

2 Park Square, Milton Park, Abingdon, Oxon OX14 4RN
711 Third Avenue, New York, NY 10017, USA

Routledge is an imprint of the Taylor & Francis Group, an informa business

First issued in paperback 2016

The Library of Congress has cataloged the printed edition as follows:
Zhang, Donia.
    Courtyard Housing and Cultural Sustainability: Theory, Practice, and Product. –
    (Design and the Built Environment Series)
    1. Architecture – Philosophy. 2. Architecture and society. 3. Sustainability.
    4. Sustainable architecture. 5. Courtyard houses. 6. Courtyard houses – China –
    History – 20th century. 7. Communities.
    I. Title II. Series
    720.1'03–dc23

Library of Congress Cataloging-in-Publication Data
Zhang, Donia, 1967-
    Courtyard Housing and Cultural Sustainability: Theory, Practice, and Product /
    by Donia Zhang.
        pages   cm. – (Design and the Built Environment)
    Includes bibliographical references and index.
    1. Courtyard houses. 2. Architecture and society. I. Title.
    NA7523.Z49 2013
    728'.31–dc23   2012038425

ISBN 978-1-4094-0503-0 (hbk)
ISBN 978-1-138-25677-4 (pbk)

Contents

List of Tables

List of Figures

Parts of picture of this book selected from *A Brief History of Chinese Residential Architecture: Cities, Houses, Gardens* © 1990 and *A History of Chinese Architecture* © 2009 China Architecture & Building Press, Baiwanzhuang, 10037, Beijing, China.

Acknowledgements

This study was made possible by the generous inputs from the residents, planners, and architects in Beijing and Suzhou, where the fieldwork was conducted. Many residents warmly welcomed me into their homes with gestures of trust. It is hoped that in return, this study will contribute positive energy to all the environments concerned.

My sincere gratitude goes to Dr Marcel Vellinga and Dr Aylin Orbasli, both at Oxford Brookes University and who have helped me refine the book structure with many constructive comments, critiques, and corrections. I also appreciate Aylin's extra effort to visit the case study sites in China.

Thanks also to Prof Michael Humphreys at Oxford Brookes University for providing valuable comments on the first three chapters. As well, my former B.Ed. facilitator, Mina Wong at Brock University, and former English language tutor, Stephen Court from King's College London, now at University and College Union, for editing and proofreading the book. The critiques made by Prof Georgia Butina Watson and Prof Ronald G. Knapp have also helped clarify certain parts of the book. I am especially grateful to Prof Ronald G. Knapp for writing the book foreword. The editors at Ashgate Publishing, Valerie Rose and David Shervington, Kirsten Weissenberg, and their desk editorial team, have contributed to improving the presentation of the book.

During the journey, the tutors at Oxford Brookes University who have taught me research methods include Dr Sue Brownill and Pete Smith. Among significant individuals who have provided me with useful resources are Chen Yiren, Zhang Fan, Li Qi, Li Hong, [Lok Tok], Prof Hu Shaoxue, Dr Zhang Jie, Prof André Casault, Dr Daniel B. Abramson, Dr Regina Lim, Dr Hom Rijal, Dr Lynne Mitchell, Dr Elizabeth Burton, Dr Carol Dair, Dr Rawiwan Oranratmanee, and Sofia C. Aleixo.

Old colleagues, friends, and family who have supported me along the way include: Brian Adams, Yueyun Cheung, Stella and Stan Rooke, Dr Sarah Stevens, Patricia Stevenson, Li Xuanzeng, Li Huanzeng, Li Yuzeng, Li Lingzeng, Li Yingzeng, Xu Dajiang, and Zheng Huiqin. I am extremely grateful to my parents Junmin Zhang and Suzan Xiuzeng Li, and sister Lydia Xiaolin Zhang, for their unconditional support.

Foreword

Ronald G. Knapp

SUNY Distinguished Professor Emeritus State University of New York at New Paltz

For many visitors to Beijing, one of their most memorable experiences is visiting a family living in a courtyard home along one of the ancient city's remaining *hutong* or alleyways. These residential *siheyuan*, which comprise adjacent quadrangles of low buildings each enclosing at least one courtyard, are surrounded by gray brick walls that provide seclusion and security while forming a template for family relationships. As a building form that has provided residences for multigenerational families in North China for centuries, *siheyuan* are quintessential courtyard houses. Elsewhere throughout China, courtyard homes are also found, but they differ in shape and size from those found in the Beijing region. This book critically examines Beijing *siheyuan*, contrasting them with the smaller courtyard dwellings in Suzhou, a historic city in China's southern Jiangnan region. Whether in Beijing or Suzhou over the past half century, these traditional housing forms have been diminished in number and elegance as China's population has escalated and modernization has brought stress to traditional neighborhoods.

Dr Donia Zhang's *Courtyard Housing and Cultural Sustainability* begins with a comprehensive examination of traditional courtyards over time. She skillfully links their form and function to important themes in Chinese philosophy: Harmony with Heaven, Harmony with Earth, Harmony with Humans, and Harmony with Self. These notions are tied to both Confucian and Daoist thought as rudimentary rules that traditionally governed daily life. Using diagrams and photographs, many of which are her own, she successfully explores eleven elements of these homes from architectural and spatial design perspectives. Since buildings only take on meaning once they are inhabited, she then leads the reader to an understanding of family organization and daily life within the rooms and open spaces of traditional homes.

With this solid information as background, Dr Zhang then embarks on what is the most original part of her book, a field-work based analysis of both renewed and new courtyard housing over the past twenty years. Focusing on six housing estates in Beijing and Suzhou, data were collected using a survey questionnaire, interviews, and personal time diary, among several informal methods including email and telephone interviews. Dr Zhang is quite frank about the difficulty of collecting this kind of data in a country such as China (and echoes my own experiences in the country). While survey data were qualitatively analyzed using SPSS, it is clear that Dr Zhang draws from her personal experience with perceptive insights in addition to secondary social science literature for context as she establishes her conclusions. These conclusions are explicit concerning the nuanced differences between Beijing and Suzhou as well as between renewed residences and those newly built.

The author is sympathetic to the intention of architects to improve living conditions for urban dwellers by creating new courtyard communal dwellings to replace those that were too dilapidated for continuing residence. Yet, in her judgment, even as the quality of the new designs and design elements was often excellent, the execution—actual construction— unfortunately was frequently unsatisfactory to the degree that surveyed residents usually focused on negatives of their residences without acknowledging any substantial positives. Dr Zhang's findings affirm also that many of the current problems with new and renewed courtyard dwellings are related to contemporary living conditions rather than the designs themselves.

This book should be read by all who are intrigued by the pace of economic change in China and its impact on the lives of those living in old courtyard dwellings or newly built ones. For general and specialized readers, it provides both an excellent introduction and a sophisticated analysis of a unique traditional housing design that has undergone dramatic changes in recent decades.

Prologue

I was born in a traditional Beijing courtyard house in 1967. When I was three, my family moved to Harbin in the far northeast of China. As my parents had to go to work during the day, I was sent back to Beijing to stay with my grandmother and aunt, who at the time, still lived in a traditional courtyard house in the Western district of the city, where I played with the neighbors' children in the courtyard almost every day of the year. We made a 'gentleman's agreement' to get up at the same time in the morning, to fly kites, to skip and dance over stretched chains of rubber bands, or to kick shuttlecocks in the courtyard. If any of us broke the 'agreement,' all of us would go and wake him/her by knocking on the door. During the winter, we would be so excited to see snowflakes falling from the sky that we could not wait to build a snowman in the courtyard, or simply throw snowballs at each other. My childhood in the courtyard was jolly, sweet, and carefree.

Before starting primary school, I went to stay with my parents in Harbin where we lived in a 4-storey communal housing with a long, common corridor on each floor. My family and four other families lived on the ground floor of the western part of the building, with all the rooms equally divided into 3 m × 5 m each. The kitchen was about one and a half times of the size of the room and was shared by five families.

There, we found new friends again! Two neighbors' children were about the same age as my sister and I, and we were all girls! We went to school and came home together every day, and then played games in the park in front of our building for the rest of the afternoon. The most memorable event to me was our 'Spring Festival Performance' organised by us three 'bigger girls' that took place in our kitchen. On New Year's Eve, all the children dressed up in new clothes, shoes, and colorful hair bands to greet the Spring Festival. At 8 o'clock on the dot, the show was 'formally' announced. All the parents, grandparents, aunts, and uncles came with their chairs to watch us. It was a great success! All the unhappy arguments among the adults about overuse common spaces seemed to have disappeared at that moment, and all the quarrels among the young girls also vanished for the time being.

When my family moved back to Beijing in 1978, I was eleven years old. We temporarily lived in my grandmother's old courtyard house for a year until my father's workplace assigned us an apartment on the 11th and very top floor of a

Figure 0.1   My memory of childhood home environment in Harbin. Drawing by Donia Zhang 2005-2012

new tower block on Qianmen Street West, about 10 minutes' walk to Tiananmen Square, and the site of the dismantled old City Walls. As it was the first time for us to see such a massive building with two elevators, we felt curious, excited, and proud! The new apartment had two rooms, 3 m × 5 m each, with a kitchen of 4 sqm, a washroom of 2.5 sqm, a corridor of 1 sqm, and a balcony of 1.5 sqm.

Life inside our apartment building was peaceful and quiet, with each family secluded from one another. This privacy actually allowed me to think about many things and form my own worldview. Sometimes, simply standing on the balcony and watching passing pedestrians in the streets was an enchanting experience. When all the houses were lit up at night, the panoramic cityscape could take my imagination far, far away. Inside these houses below, each family told its own story, and each story unfolded its own bitter-sweetness. Thus, to this day, I consider the house as the backdrop for the people who lived out their stories.

In 1990, my family moved to a bigger apartment on the 12th floor of a tower block on Art Gallery Rear Street in the northeast of the Forbidden City, within walking distance to Beihai and Jingshan parks in Beijing. The lifestyle there was more or less the same, apart from the fact that we could invite more relatives for Chinese New Year dinner or family members' birthday celebrations.

Sometime after emigrating from China to Canada in late 1990s, my family and I have lived in a Western-style suburban house in Ontario. The house faces a community park – King's College Park, reminiscent of a Chinese courtyard, except that the neighbors seldom communicate with one another.

To conduct fieldwork for this research, I arrived in Beijing in September 2007. Twelve years after being away, I was surprised to discover things that had remained constant and those that had changed. Beijing is now entirely a new city with only bits and pieces of old buildings left. Walking down memory lane to

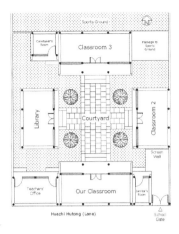

Figure 0.2   My memory of Liuhai Primary School in courtyard style in Beijing. Drawing by Donia Zhang 2010

my Liuhai Primary School (刘海, the name of a Daoist immortal) that was in the courtyard style, I was disappointed to find the school demolished and rebuilt as a 4-storey building occupied by another institution. The school that had appeared in my dreams no longer exists, but I still hold fond memories of it. It was there that I first heard from my arithmetic teacher about UFOs that sparked my curiosity of the universe; it was also there that I felt safe playing with other pupils in the courtyard until the bell rang. Given only remembrance, I have sketched out the plan of the school from memory.

Cultural constancy and change is an experience that all the Chinese people readily accept throughout life. Ordinary citizens in contemporary China do not care much about the old courtyard houses; many still see them as a symbol of the old society and shelter for lower-class citizens: they represent backwardness. When I told Beijingers about my research topic, their reaction was always the same: "*siheyuan* (courtyard house)? Who would want to live there anymore? People living in them cannot move out soon enough. Who cares about keeping them?" Many people there felt that I was wasting time studying such an unwelcome topic.

Sir Winston Churchill said: "A pessimist sees the difficulty in every opportunity; an optimist sees the opportunity in every difficulty." *Feng Shui* theory advises that whatever task we undertake we need to acquire "the right time, the right place, and the right people" (天时, 地利, 人和). In other words, to succeed, we must be closely in tune with both nature and culture. More than a decade later, the political milieu in China has permitted me to conduct such research, with the study's projects being available and accessible and the associated individuals cooperative. Thus, these conditions have offered me the possibility to complete the study.

DONIA ZHANG
January 2012

Chapter 1

Introduction
Lost Link between Heaven and Earth

Heaven is my father and Earth is my mother,
and even such a small creature as I find an intimate place in their midst.
Therefore that which fills the universe I regard as my body and
that which directs the universe I consider as my nature.
All people are my brothers and sisters, and all things are my companions.
         – Zhang Zai, 1020-1078, *Western Inscription*; cited in Chan, 1969, p. 497

Traditional Courtyard Houses

The courtyard house is one of the oldest dwelling typology, spanning at least 5,000 years and occurring in distinctive forms in many parts of the world across climates and cultures, such as China, India, the Middle East and Mediterranean regions, North Africa, ancient Greece and Rome, Spain, and Latin-Hispanic America (Blaser, 1985, 1995; Edwards *et al.*, 2006; Land, 2006, p. 235; Pfeifer and Brauneck, 2008; Polyzoides *et al.*, 1982/1992; Rabbat, 2010; Reynolds, 2002).

Archaeological excavations unearthed the earliest courtyard house in China during the Middle Neolithic period, represented by the Yangshao culture (5,000-3,000 BCE) (Liu, 2002a). The ancient Chinese favored this housing form because enclosing walls helped maximize household privacy and protection from wind, noise, dust, and other threats; and the courtyard offered light, air, and views, as well as acting as a family activity space when weather permitted (Knapp, 2005a; Wang, 1999).

There was a distinctive variety of traditional courtyard houses due to China's wide-ranging climates, 56 ethnic minority groups, and notable linguistic and regional diversity even among the Han majority. Hence, traditional Chinese courtyard houses were grouped as northern, southern, and western types according to their geographic locations in relation to the Yangzi River (Knapp, 2000, p. 2; figure 1.1). Chinese city planning and courtyard houses first emerged in the north and eventually appeared in the south due to migration (Knapp, 2000, p. 223; Wu, 1968). Cultural diffusion began during the Ming dynasty (1368-1644) when core populations were encouraged to move to other parts of China to enhance their growth; most settlers built new residences in the style of their original homes (Sun, 2002). This confluence of building patterns demonstrates the layers and veins of acculturation within China over the centuries.

Figure 1.1   **Map of China showing the northern, southern, and western divisions with locations of Beijing and Suzhou as the case study cities. Source: adapted from Knapp, 2000, p. 2 © 2005 University of Hawaii Press. Reprinted with permission**

Chinese builders throughout the country have historically favored a number of conventional building plans and structural principles such as bilateral symmetry, axiality, hierarchy, and enclosure. Courtyards or lightwells (天井 *tianjing*, or 'skywells') were important features in the layout of a fully built Chinese house. Philosophically, the courtyards acted as links between Heaven and Earth, as during the Han dynasty (c.206 BCE-220 CE), the Chinese regarded Heaven and Earth as a macrocosm and the human body a microcosm to reflect the universe (Chang, 1986, p. 200); offering sacrifices to Heaven and Earth in courtyards was considered crucial to bringing good fortune (Flath, 2005, pp. 332-334).

Across China, four basic types (and their sub-types) of traditional courtyard houses are found:

1. 1-storey square/rectangular-shaped courtyard houses (*siheyuan*) in the northeast such as in Beijing, Hebei, and Shandong; 1-2-storey rectangular courtyard houses that are longer in the north-south direction to maximize sunlight in the north such as in Shaanxi and Shanxi;

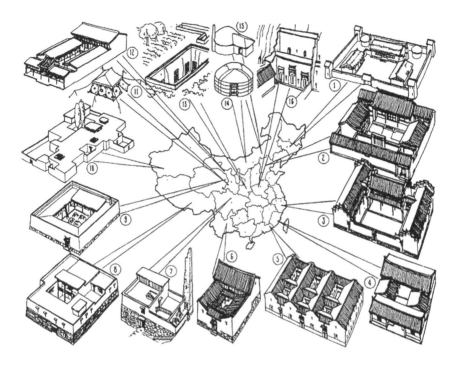

Figure 1.2 **House types across China showing the courtyard as a common feature. Source: Drawing by Fu Xinian in Liu, 1990, p. 206**

2. 2-3-storey courtyard/lightwell houses that are longer in the east-west direction to filter out summer's hot sun in such southern regions as Jiangsu; 2-storey inverted U-shaped courtyard/lightwell houses (*sanheyuan* and *yikeyin*) in central and southern provinces such as Sichuan, Hunan, Anhui, Zhejiang, Fujian, Guangdong, Guanxi, Yunnan, and Taiwan;
3. 3-4-storey fortresses in circular, elliptical, or octagonal structures (*tulou*) that house groups or entire clans in southern regions such as Fujian, Guangdong, and Jiangxi;
4. Cave-like sunken, earthen, or subterranean courtyard houses in northern and northwest regions such as Shanxi, Shaanxi, and Henan (Knapp, 2000, 2005a; Liang, 1998; Wu, 1999).

Beijing *siheyuan* are commonly regarded as the most outstanding examples of traditional northern Chinese courtyard houses.

Traditional Suzhou courtyard houses are representative of the southern type, generally with smaller courtyards and gardens to admit less sunlight due to their hot summers.

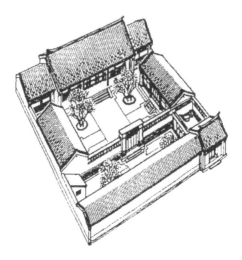

Figure 1.3   A two-courtyard house of Beijing. Source: Liu, 1990, p. 210

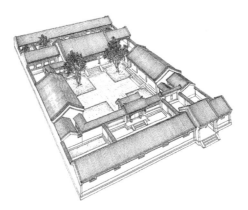

Figure 1.4   A typical or standard three-courtyard house of Beijing. Source: Ma, 1999, p. 7

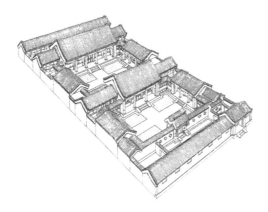

Figure 1.5   A relatively large four-courtyard house of Beijing. Source: Ma, 1999, p. 19

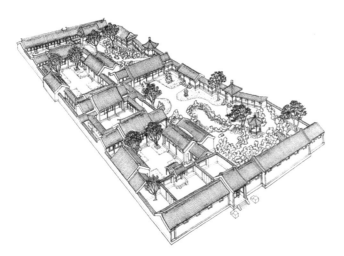

Figure 1.6   A five-courtyard house of Beijing with gardens. Source: Ma, 1999, p. 227

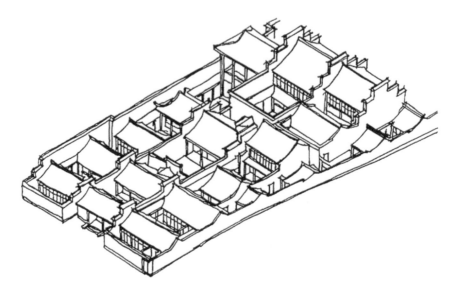

Figure 1.7   A large courtyard house compound in Suzhou. Source: Wu, 1991c, p. 58

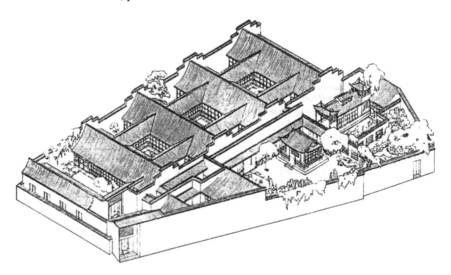

Figure 1.8   A typical southern type of courtyard house complex with small courtyards and gardens in Yangzhou, China. Source: Schinz, 1989, p. 58, Urbanisierung d. Erde, Vol. 7, Borntraeger Science Publishers, www.borntraeger-cramer.de

Decline of Courtyard Houses

Traditional Chinese courtyard houses have undergone changes over the years. The Qing dynasty (1644-1911) was the peak period for the development of courtyard houses. However, British and French attacks from the 1840s, the 1911 *Xinhai* Revolution, and Japan's two invasions during 1894-1895 and 1937-1945 forced a massive population move from the suburbs into the cities, resulting in a severe urban housing shortage and rendering most citizens' livelihood increasingly more difficult. With rapidly increasing prices, many households started renting out rooms in their courtyard houses to help with living expenses. Consequently, the originally single-extended-family homes became multifamily compounds, and the very nature of traditional Chinese courtyard houses changed (Bai, 2007; Ma, 1999).

With the establishment of People's Republic of China in 1949, the country's population expanded rapidly. Subsequently, the evolved 4-5-family courtyard houses increased capacity to shelter over 10 families upon their children's marriages. New kitchens, storage spaces, and even bedrooms built into the courtyards turned the ordered enclosures into disordered quarters. These changes made it difficult for the courtyard houses to maintain their original charm, mystique, tranquillity, and cosiness (Bai, 2007; Ma, 1999; Zhu, Huang, and Zhang, 2000, p. 8).

During the 'Cultural Revolution' (1966-1976), traditional Chinese courtyard houses sustained the most severe damages. Systematic destruction was launched in 1966 to demolish the 'Four Olds' (old ideas, old culture, old customs, and old habits). Red Guards chiselled out uncountable amounts of invaluable artwork of brick, wood, and stone carvings; they also ruined colorful decorations on courtyard houses, after which, the few surviving houses were either plastered over, repainted, or simply left in their own wreckage. This devastation led not only to demolished courtyard houses, but an irreversible, fatal assault on Chinese art and architecture (Knapp, 2005a; Ma, 1999; J. Wang, 2003, p. 335).

More damage was still on the way. In March 1969, at the height of the Sino-Soviet split, a border conflict occurred in Zhenbao Island that led to a nation-wide 'Excavation Movement'[1] on the Chinese side to prepare for Russian attacks, but which further destroyed traditional courtyard houses' structures. In Beijing, this movement was eclipsed by the 1976 Tangshan Earthquake that was like 'frost on snow.' To provide refuge, many 'anti-quake sheds' were added to already overcrowded courtyards. These temporary shelters became permanent and distorted Beijing *siheyuan* beyond recognition (Bai, 2007; Ma, 1999; figure 1.9).

Since the 1990s, China's rapid economic growth coupled with an unprecedented level of real estate development has resulted in the almost wholesale destruction

---

1   The movement came with a slogan raised by Mao Zedong in 1972: 'dig tunnels deep, store grain everywhere, and never seek hegemony,' based on 12 original words from Liu Bang's guiding principle during the Three Kingdoms (220-280) when his power was weak: 'dig tunnels deep, store grain everywhere, build walls high, and declare kingship slow.'

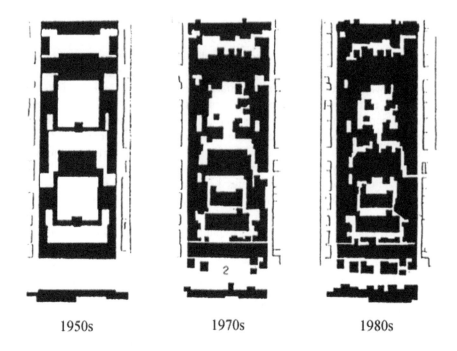

| 1950s | 1970s | 1980s |

Figure 1.9 **Decline of a courtyard house compound in Beijing from early 1950s – 1970s – 1980s. Source: Zhu and Fu, 1988**

of traditional courtyard houses. For example, until 1949, Beijing was a completely traditional courtyard city. In the early 1950s, Beijing's inner city had 11 sqkm of single-storey *siheyuan*, of which only 5-6 percent was dilapidated. At the end of the 1980s, most parts of inner Beijing were still occupied by *siheyuan* of various qualities. However, in 1990, the inner city had total single-storey *siheyuan* of 21.42 sqkm, of which almost 50 percent was decaying. The increased floor space from 11 sqkm to 21.42 sqkm was due to the proliferation of improvised extensions, an indication that not much courtyard was left. Between 1990 and 1999, a total of 4.2 sqkm of Beijing *siheyuan* were demolished, with its areas shrunk from 17 sqkm to 3 sqkm between the early 1950s and 2005 (Tan, 1998; Yuan, 2005). Today, the few well-preserved *hutong* ('lane') with refurbished *siheyuan* serve only high officials and those who can afford such homes (Trapp, 2002?, p. 5; Zheng, 2005).

Table 1.1   Extent of survival of *siheyuan* and *hutong* in inner Beijing

Year	*Siheyuan* (courtyard house)	*Hutong*
1949	100 percent (of 6,200 hectares of inner-city land area)	7000
1990	1.9 percent (805 courtyard houses in the conservation areas)	3900
2003	1.5 percent (658 courtyard houses in the conservation areas)	1570
2004	1.3 percent (539 courtyard houses in the conservation areas)	1200

Sources: Abramson, 2001; *Beijing City Planning Chart*, 2007; Collins, 2005; Kong, 2004; Ornelas, 2006

The massive destruction of traditional Chinese courtyard houses was followed by housing redevelopment to alleviate acute housing shortage. From the late 1950s to the late 1970s, the Chinese government constructed large numbers of new residential quarters of 4-5-storey parallel 'Socialist Super Blocks' influenced mainly by the Soviet Union (Dong, 1987; Gaubatz, 1995, 1999). Between 1974 and 1986, the Beijing Municipal Government built around 7 sqkm of new housing in the inner city which accounted for 70 percent of the city's total housing redevelopment since 1949 (Wu, 1999), most of which consisted of residential tower blocks of over 10 storeys comprised of individual apartments. By the end of 1996, new housing projects numbered over 200, covering 22 sqkm of inner Beijing (Tan, 1997, 1998). In Suzhou, between 1994 and 1996, nearly 3 million sqm of new housing were added to the housing stock each year (Zhu, Huang, and Zhang, 2000, p. 3).

Such a large scale housing redevelopment has had a major impact on the historic cities' overall structure, land-use pattern, and cultural sites. Although each new apartment has the advantage of privacy and facilities, not to be underestimated when compared with the old, dilapidated courtyard houses from which many of these residents came, and the mid- and high-rises had eased some desperate housing demands, most of them did not include a *courtyard* – the traditionally important link between Heaven and Earth was lost. Chinese urban areas are extending and transforming at such phenomenal speed that the intangible value of traditional dwelling culture runs the danger of uncontrolled elimination (Acharya, 2005; Collins, 2005; Chatfield-Taylor, 1981; Entwurf, 2005; Gaubatz, 1995, 1999; Mann, 1984; Marquand, 2001; Nilsson, 1998; Ornelas, 2006; Reuber, 1998; Schell, 1995; Spalding, n.d.; Tung, 2003; Van Elzen, 2010; W. Wang, 2006; Watts, 2007; Whitehand and Gu, 2006; Wood, 1987).

As early as the Han dynasty (206 BCE-220 CE), China had already launched debates concerning high-rise buildings with the concluding remark that they were "too far to harmonize Heaven and Earth, therefore drop the idea"[2] (Luó, 1998; Luò, 2006, p. 109; J. Wang, 2003, p. 146; my translation). A growing body of

---

2   远天地之和也，故人弗为。

literature provides evidence for the unnatural and harmful impacts of tower blocks not only on Chinese cultural landscapes, but also on the wellbeing of occupants, particularly that of the elderly and children (Bosselmann *et al.*, 1984; Ekblad and Werne, 1990a, 1990b; Ekblad *et al.*, 1992; Lu, 2004, p. 333; Stewart, 1970; Tan, 1997; J. Wang, 2003, p. 145; Wu, 1999; Yan and Marans, 1995).

Numerous housing studies suggest that despite traditional courtyard houses' physical deteriorations and poor living conditions (He, 1990; Lu and He, 2004; Wu, 1999, p. 112), Beijing residents prefer renovating traditional courtyard houses to relocating to high-rise apartments (Alexander *et al.*, 2003; Hou, 2006; Song, 1988; Tan, 1994, 1997; Yan and Marans, 1995; Zhao, 2000). It is found that traditional courtyard houses foster better social relations than new housing forms, and that people living in traditional courtyard houses seem more satisfied with life because social connectedness among the residents plays a more important role than housing conditions in relationship to stress, life satisfaction, and health (Bai, 2007, pp. 73-76; Ekblad and Werne, 1990a, 1990b; Ekblad *et al.*, 1992; Hou, 2006; Yan and Marans, 1995; Zhao, 2000, pp. 106-107).

A number of research findings also indicate that residents conduct social and cultural activities in traditional courtyards much more often than in the outdoor spaces of mid- or high-rise apartment buildings, because traditional courtyard houses contain a clearly defined hierarchy of spatial transition from public to private that the mid- or high-rise modern housing outdoor spaces lack. Furthermore, new housing's outdoor spaces are frequently perceived as potential breeding ground for criminal activities (Han, 2001; Kanazawa and Che, 2002; Qi and Yamashita, 2004; Zhang, 2005). Thus, understanding the role of the social environment in the wellbeing of occupants of different housing types is important.

Historic preservation only began to be taken seriously in China since the 1980s (Abramson, 2001, 2007; Ma, 1999; Together Foundation and UNCHS, 2002) to reinforce the courtyard house protection policy so that earlier blunders would be somewhat controlled. Nevertheless, destruction occurred periodically, and the most devastating case is the demolition of the courtyard home where revered Chinese architects Liang Sicheng and his wife Lin Huiyin once worked (Branigan, 2012).

A common criticism of courtyard houses is that they take too much land (Ma, 1999). However, several research studies demonstrate that courtyard form is in fact efficient in land-use. For example, Shang and Yang (1982) conducted a comparative study of the relationship between open and closed spaces on building lots of the same size in China and the USA (figure 1.10), and found that the traditional Chinese courtyard house (left) had a ratio of 7:3 between the closed and open areas, with the courtyard space encircled by surrounding structures still widely available for use. This form was the opposite of a typical American house (middle), which generally sits at the center of its lot. If discarding the courtyard and copying the American residential layout to create the same amount of building space as in a traditional Chinese courtyard house, the yard would be left with a fringe and become useless (right) (Knapp, 2005c, pp. 58-59).

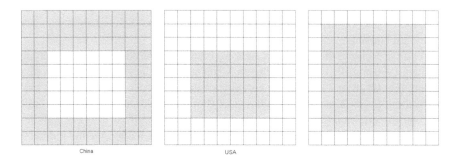

Figure 1.10 Comparison of the relationship between open and closed spaces on building lots of the same size in China and the USA. Source: Drawing by Donia Zhang after Shang and Yang, 1982, p. 56

This result concurs with the findings by Martin and March (1972) from the Martin Centre of Architectural and Urban Studies at Cambridge University that, using the same area of land, the courtyard form (or perimeter block) is the most economic land use. This also echoes Tsinghua University's research result that it is possible to achieve a high ratio of floor area to land area with a 2-3-storey courtyard form, even in a 9 m height restricted zone (Wu, 1999, pp. 120-124). My M.A. research discovery is consistent with this claim (Zhang, 1994, 2006, 2009/2010/2011).

Research into New Courtyard Housing

New courtyard housing has been built in recent years in China which reflects people's nostalgia for traditional courtyard houses, a new appreciation for Chinese architectural heritage, and a response to the cultural crisis of modernity (Ma, 1999; Wu, 1999, pp. 78-80; Zhao, 2000, p. 79). However, the projects were few and were initiated by the municipal governments (Jin *et al.*, 2004).

This study investigates new courtyard housing in two Chinese cities: Beijing and Suzhou. Both cities have been chosen because they have followed the city planning principles set in the *Record of Trades* in *Rituals of Zhou* and *Feng Shui* theory. Their traditional courtyard houses embedded in their urban fabric are representative of traditional Chinese urban culture despite their climatic differences. Beijing is a northern Chinese city with a rich history of 3,000 years and as China's capital for 800 years; its famous *siheyuan* with strict axial, symmetrical, and hierarchical planning embody the Confucian ideal of 'harmony in social relationships.' Suzhou is a southern Chinese city with a prosperous history of 2,500 years, and was a regional capital renowned also for its private gardens enclosed within courtyard house compounds, whose spontaneous layouts reflect the Daoist principle of 'harmony with nature.' They were thought to offer a good comparison of their traditional courtyard use and the contemporary new courtyard housing.

In inner Beijing, the first new courtyard housing is the Juer Hutong prototype (built in 1990 and 1994) in the Nanluogu Xiang area, and the second is the Nanchizi experiment (built in 2003) on the east side of the Forbidden City. In the old city of Suzhou, projects include Tongfangyuan (built in 1996) and Shilinyuan (built in 2000) by the Lion Grove Garden, and Jiaanbieyuan (built in 1998) in walking distance of the Master-of-Nets Garden and the Canglang Pavilion, among others. This housing redevelopment endeavours to reinterpret classical Chinese courtyard houses while resettling multiple families in 2-storey town/terraced houses or 2-4-storey walk-up apartments surrounding a communal courtyard.

Previous research findings showed that residents responded favorably to the new courtyards. At Beijing Juer Hutong new courtyard housing, 46 percent of residents felt they were home when entering the communal courtyard (Liu Wenjie, unpublished report, 1992; Wu, 1999). At Suzhou Tongfangyuan new courtyard-garden housing, 57 percent of respondents felt a sense of familiarity after entering their estate's gate (Jin *et al.*, unpublished report, 2004). This reaction indicates that the communal courtyards have enlarged residents' perception of 'home' to include their courtyard, which may also expand their idea of home to include the natural world and other cohabitants. Thus, the new courtyard may foster residents' environmental and social consciousness.

Traditional Chinese courtyard houses are declining, and new ones are being designed and built. What is not known, however, is that to what extent the new courtyard housing is a continuation of the old ones that allows traditional ideas and activities to be sustained; in other words, is the new courtyard housing sustainable from a cultural perspective?

Aims and Objectives

The aims of this research are to investigate residents' perceptions and experiences of living in the renewed/new courtyard housing in Beijing and Suzhou, to see whether the housing is culturally sustainable, and whether it facilitates residents' traditional cultural expressions. As such, the objectives are:

1. To theorize the concept of cultural sustainability;
2. To identify key themes in Chinese philosophy that have influenced city planning and classical courtyard house design;
3. To investigate which design elements, if any, in courtyard housing help to promote cultural sustainability, and why?
4. To theorize culturally sustainable architecture in China.

Chinese philosophy is incorporated into the study to bring attention to the poetic and intangible aspects of architecture, and to examine how they traditionally manifest themselves in building forms, space use, orientation, and similar contexts.

This information becomes the benchmark against which change and continuity will be measured.

Through a literature review, four key themes in Chinese philosophy that have influenced imperial city planning and classical courtyard house design have been identified: harmony with heaven, harmony with earth, harmony with humans, and harmony with self, along with form and environmental quality, space and construction quality, matters of social cohesion, and time and cultural activities, to establish an ordering principle for structuring the empirical findings (Chapters 7-10). These four aspects comprise the four cornerstones of culturally sustainable architecture in China (Chapter 11).

Significance of the Study

This study defines cultural sustainability as the adaptation and transmission of the beneficial parts in a nation's material (tangible) and immaterial/spiritual (intangible) culture that are conducive to the development of their present and future generations. It encompasses such notions as cultural vitality, cultural diversity, and cultural activities. Culture as the fourth pillar of sustainable development was popularized by Jon Hawkes' book *The Fourth Pillar of Sustainability: Culture's Essential Role in Public Planning* (2001), but it was neither fully explored nor widely recognized. It is hoped that this study will enhance our understanding of cultural sustainability, which is an original and significant research problem.

There is an obvious lack of human behavioral study of the new courtyard housing, including those aspects of it that can be adapted to contemporary living (Özkan, 2006, p. xvii; Zhao, 2000), this research attempts to fill the knowledge gap. There is also little previous study on Suzhou housing development in past decades compared with Beijing, the study hopes to reduce that disparity.

Research Methodology

This research is a collective case study using quantitative and qualitative methods, and taking the six approaches to vernacular architecture as outlined in Paul Oliver (1997a): architectural, historical, aesthetic, spatial, anthropological, and behavioral. An extensive historical research was undertaken, with scriptures extracted from Chinese philosophical books presented throughout the work. These efforts aim at establishing the context and its textual evidence, and to reveal the cultural forces that have shaped imperial Chinese capital cities and classical courtyard houses. A detailed literature review on Beijing and Suzhou's traditional courtyard houses was conducted, of which more than 50 percent of the texts were translated from Chinese sources.

The empirical fieldwork was carried out in six housing estates in Beijing and Suzhou, three cases in each city. A multiplicity of data was collected to triangulate the findings so that they could provide checks against the weak points in each: field surveys, time diaries, email/telephone interviews, photos, architectural drawings, planning documents, conversations and on-site observation notes, and real estate brochures and fliers. Due to time and money constraints, the field study was conducted in one trip to China in 2007.

Limitations and Delimitations of the Study

This study considers architecture as a cultural artefact and evaluates both archi-cultural[3] and socio-cultural aspects of the renewed/new courtyard housing in China. Hence, the cultural indicators identified, such as the four key themes in Chinese philosophy, *Feng Shui* theory, and cultural activities and festivities, will pose limited generalizations for other cultures.

Architectural acculturation has resulted in some creative courtyard forms in the style of courtyard villas built across China since the 2000s, but because these new courtyard houses are not social housing and it was not feasible to investigate them, they are not included here. Also, due to less existing literature on Suzhou's recent housing (re)development than that of Beijing, there is a difference in coverage and some case descriptions (Chapter 6).

Structure of the Following Chapters

Chapter 2 reviews the concept of sustainable development and theorizes cultural sustainability while making a link to courtyard housing. Chapter 3 identifies four key themes in Chinese philosophy and explains how they influenced imperial city planning and classical courtyard house design. Chapter 4 reviews the architecture of Chinese classical courtyard houses to build the research context. Chapter 5 traces the socio-culture of classical Chinese courtyard houses for comparison with that in the new courtyard housing. Chapter 6 examines Chinese historic preservation policies and introduces the six case studies. Chapter 7 evaluates the architectural and environmental aspects of the renewed/new courtyard housing. Chapter 8 examines the spatial and constructional aspects of them. Chapter 9 investigates the social dimension of the cases. Chapter 10 reveals the cultural beliefs and behavioral patterns of residents in the renewed/new courtyard housing. Chapter 11 synthesizes the findings and theorizes culturally sustainable architecture in China. To conclude the study, Chapter 12 emphasizes the contribution to the tree of knowledge and suggests the creation of new courtyard garden houses.

---

3   The word 'archi-cultural' is my derivative of Nancy Berliner's term 'archiculture' (see Knapp, 2005b, p. 9) to denote the intrinsic connection between culture and architecture.

PART I

Cultural Sustainability

Chapter 2

Four Pillars of Sustainable Development: Courtyard Housing and Cultural Sustainability

The way of Heaven and Earth may be completely declared
in one sentence: they are without any doubleness,
and so they produce things in a manner that is unfathomable.
The way of Heaven and Earth is large and substantial,
high and brilliant, far-reaching and long-enduring.
— Confucius, 551-479 BCE, *Doctrine of the Mean*, Chapter 26

This chapter theorizes the concept of cultural sustainability based on a literature review, because cultural sustainability is a part of the theoretical framework of the study which considers architecture as a cultural artefact. It attempts to explicate how architecture is intricately connected to environmental, economic, social, cultural, and political factors, with a particular emphasis on cultural changes in China. The chapter consists of five sections: environmental responsibility, economic viability, social equity, cultural vitality, and political support, with cultural threads running through the literature on sustainable development.

The word 'sustainability' derived from the Latin *sustinere* (*tenere*, to hold; *sus*, up). Dictionaries provide more than 10 meanings for 'sustain,' the main ones being to 'maintain,' 'support,' or 'endure' (Dictionary.com, 2009; Onions, 1964, p. 2095). Since the 1980s, sustainability has been used more in the sense of human sustainability on planet earth, resulting in the most widely quoted definition of sustainable development, that of the Brundtland report *Our Common Future* (1987) published by the United Nations World Commission on Environment and Development: "sustainable development is development that meets the needs of the present without compromising the ability of future generations to meet their own needs" (p. 8). Since then, the concept has evolved and was presented in the Rio Declaration on Environment and Development, also known as the Rio Earth Summit (UNCED, 1992), that states, "The right to development must be fulfilled so as to equitably meet developmental and environmental needs of present and future generations" (UNCED, 1992, article 3).

However, the UN definition is not universally accepted and has undergone various interpretations (EurActiv, 2004; International Institute for Sustainable Development, 2009; Kates, Parris, and Leiserowitz, 2005). The 2005 World Summit acknowledges that sustainability requires the reconciliation of environmental, social, and economic demands known as the 'three pillars': environmental responsibility, economic viability, and social equity (Bell, 2003; OECD, 2001; United Nations General Assembly, 2005). The United Cities and Local Governments (2006a, 2006b, 2006c, 2006d, 2009, 2010) share the view that culture is the 'fourth pillar' of sustainable development, a notion popularized by Jon Hawkes' book *The Fourth Pillar of Sustainability: Culture's Essential Role in Public Planning* (2001) to square the sustainability triangle. Sustainability has now been commonly recognized as having 'four pillars,' including cultural vitality.

Environmental Responsibility

The environments we inhabit indeed affect human development (Diamond, 1999), and a healthy environment provides the necessary external conditions for sustaining a culture. Environmental responsibility calls for an ecological balance (Hawkes, 2001, p. 25) that includes an efficient use of resources. Sustainable development has traditionally been centered on the issue of ecological degradation and environmental protection perceived mainly through the lenses of environmentalists. As the concept evolves, increasing emphasis has been laid on its interconnection to social, cultural, and economic dimensions (Kadekodi, 1992; Nurse, 2006).

In his book, *The Nature of Design* (2002), David Orr advocates ecological design which he defines as "an art by which we aim to restore and maintain the wholeness of the entire fabric of life increasingly fragmented by specialization, scientific reductionism, and bureaucratic division" (p. 29). To him, ecological design describes the assembly of schemes and technologies by which societies use nature to construct culture to meet human needs (Orr, 2002, p. 14). As such, ecological design is a large concept that connects the natural and social sciences and applied arts (architecture) to ethics, politics, and economics (Orr, 2002, p. 4). Orr argues that the ultimate purpose of ecological design is to build a sustainable world which honors human dignity, nourishes the human spirit, and respects the human need for roots and connection (Orr, 2002, p. 30), and that the essence of ecological design is neither efficiency nor productivity, but health (Orr, 2002, p. 29). Orr also argues that there are important lessons for humans to relearn about the arts of longevity, now called 'sustainability,' from earlier cultures (p. 11). Orr (2002, p. 180) further contends that ecological design should be focused on giving priority to the wisdom of the design intentions rather than the cleverness of the means with the four underlining principles:

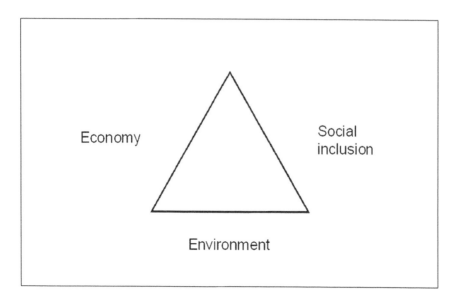

Figure 2.1 **The earlier 3-pillar sustainable development framework. Source: after UNESCO, 2006**

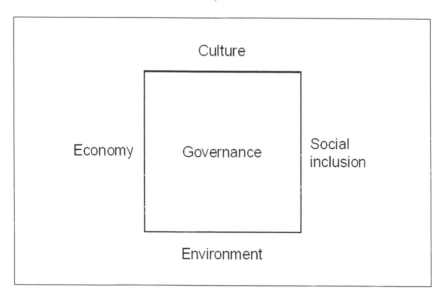

Figure 2.2 **The accepted 4-pillar sustainable development framework. Source: after UNESCO, 2006**

1. Ecological design is a community process that aims to increase local resilience by building connections between people, between people and the ecology of their places, and between people and their history; it restores connections at the community level;
2. Ecological design takes time seriously by placing limits on the velocity of materials, transportation, money, and information. The old truism 'haste makes waste' makes intuitively good ecological design sense;
3. Ecological design eliminates the concept of waste and transforms our relationship to the material world because ecologically, there is no such thing as waste; all materials are 'food' for other processes. Ecological design is the art of linking materials in cycles and thereby preventing problems of careless use and disposal;
4. Ecological design at all levels has to do with system structure, not rates of change. The ecosystem operates by the laws of thermodynamics and processes of evolution and ecology that are played out over the long term. It challenges the politics of regarding economic growth as the supreme value (Hester, 2006; Higgs, 2003; Hough, 1995/2004; Lyle, 1994; Orr, 2002).

Previous housing studies show that traditional courtyard housing encourages optimum environmental performance, such as obtaining more sunlight, air, and thermal comfort than other housing forms (see, for example, Edwards *et al.*, 2006; Ettouney, 1975; Ford, Lau, and Zhang, 2006; Heidari, 2000; Hsin, 2004; Kim, 2001; Rajapaksha, Nagai, and Okumiya, 2002; Reynolds, 2002). Hence, arguably, the traditional courtyard housing form is an ecological design. This study will consider whether the ecological design aspects of traditional courtyard houses are translated into contemporary courtyard housing projects.

The Charter for the Conservation of Historic Towns and Urban Areas (ICOMOS, 1987) and the Charter on the Built Vernacular Heritage (ICOMOS, 1999), among others, provide a set of principles to endorse the protection of the ecological and architectural heritage of humanity. Thus, the safeguard of cultural environments is the first priority in our aim to achieve sustainability, and courtyard housing may contribute to it.

Economic Viability

Economic viability refers to whether a given solution is economically possible for the group of people it is serving. It reflects the need to achieve a balance between the costs and benefits of economic activity within the limits of the carrying capacity of the environment; economic progress should not be made at the expense of intergenerational inequity, and resources should not be exploited to compromise their regenerative ability (Hawkes, 2001; Munro, 1995; Nurse, 2006).

Connected with and complementary to economic viability is cultural capital. Cultural capital refers to nonfinancial assets that involve educational, social, and intellectual knowledge. The concept was first articulated in *Cultural Reproduction and Social Reproduction* (1973) by Pierre Bourdieu, for whom cultural capital acts as a social relation within a system of exchange that includes the accumulated cultural knowledge that confers power and status. For example, educational attainment and/or the ability to converse knowledgeably about high culture have traditionally been a form of cultural capital associated with the middle class.

Cultural capital is distinguished from social capital (whom you know) and economic capital (wealth). According to David Throsby (2008), an item of cultural capital can be defined as an asset that gives rise to cultural value in addition to whatever economic value it might possess. Cultural capital defined in this way may exist in intangible form as ideas, beliefs, traditions, and practices, or in tangible form as buildings, sites, artworks, and artefacts. A heritage building may have some commercial value, but its true value to individuals or a community is likely to have aesthetic, symbolic, spiritual, or other attributes that may transcend or lie outside the economic measure, and can be regarded as the building's cultural capital (UNESCO-UIS, 2009, p. 49).

Throsby (2008) argues that there are striking similarities between economic and cultural capital, and that a healthy and flourishing environment for creative artists and individuals is necessary to support the more commercial operations of the cultural sector (p. 5), and to contribute to the economic empowerment of a nation. However, Wheelwright (2000) criticizes Throsby's commoditization of sustainability and his mathematical calculation of cultural capital entirely based on the method of evaluating economic capital.

Arts and cultural facilities bring economic development and revitalization to neighborhoods, according to a study by Ryerson University's Center for the Study of Commercial Activity (Creative City Network of Canada, 2004b, p. 1). Architect Bing Thom argues that our economic wellbeing is based on our ability to celebrate the cultural identity of our cities (Creative City Network of Canada, 2004a). Strong, creative communities may attract business and industry, and may subsequently bring employment opportunities and increase the wealth of the community (Creative City Network of Canada, 2004c, p. 3).

However, the concepts of 'creative city' by Charles Landry (2000) and 'creative class' by Richard Florida (2002, 2008), attempting to promote local economies using cultural activities, have met supporters and critics alike. For example, Maria Stanborough (2011) observes that these ideas have positively made local governments begin to seek "the best of what their community has to offer in the pursuit of cultural identity and economic prosperity" (p. 96).

Economic viability generates revenues essential to sustain a culture. While the media frequently report that China has become the world's second largest economy after the United States, this perception has contributed to the promotion of traditional Chinese culture. For example, having founded 316 Confucius Institutes and 337 Confucius Classrooms in 94 countries and regions worldwide by 2010, it plans to launch 1,000 Confucius Institutes by 2020 globally (Confucius

Institute Online, 2009; Xinhua News Agency, 2010-07-13). In time, this impetus may revive a cultural tradition.

In the TV documentary series *The Rise of the Great Nations* (大国崛起) by China Central Television (CCTV, 2006), Yale University Professor Paul Kennedy argues that what makes a great nation in today's world may ultimately be a cultural attribute. In the same program, former President of France, Giscard d'Estaing asserts that the rise of a great civilization possessing a great cultural tradition and wisdom may be conducive to the progress of the whole world.

At a conference held in Beijing in 2000 entitled *China-Cultural Heritage Management and Urban Development* organized by UNESCO World Bank, the World Tourism Organization estimated that in the year 2020, China will be the leading destination with 130 million tourists. Cultural heritage is an economic source and historic towns and cities are an asset for the future (du Cros *et al.*, 2005). The mayor of Beijing, Liu Qi, clearly pointed out: "For the economic development of Beijing, its historic and cultural heritage is not a burden but is rare economic, cultural, and social resource, and the utilization of these resources is an important aspect of modernization" (Cao, 2001; my translation). During the 2008 Beijing Olympics, 598 households with 726 rooms in renovated *siheyuan* (courtyard houses) opened to tourists as lodgings so that they could savor some of Beijing's lingering charm and old lifestyle (Du, 2008). Thus, regeneration not only offers material improvement of citizens' living conditions, but also generates revenues to help maintain traditional courtyard houses for ordinary citizens.

Social Equity

Social equity refers to matters such as justice, participation, cohesion, and welfare in a society; it relates to the maintenance of intangible social values and norms of a society that influence its social relations; it also aims to meet the basic human needs such as food, clothing, shelter, and equal distribution of these goods (Hawkes, 2001; Nurse, 2006). As Creative City Network of Canada (2007b) states:

> Sustainability is fundamentally about adapting to a new ethic of living on the planet and creating a more equitable and just society through the fair distribution of social goods and resources. Sustainable development questions consumption-based lifestyles and decision-making processes that are based solely upon economic efficiency, but its ethical underpinnings go beyond obligation to the environment and the economy – it is a holistic and creative process towards which we must constantly strive. (p. 3)

The social dimension of sustainability necessitates sustainable community development (Duxbury and Gillette, 2007). It has been suggested that a community's vitality is closely related to its cultural participation, expression, exchange, and celebration, and that the contribution of culture to building lively cities and communities where people desire to live, work, and visit has a major role to play in supporting social, economic, and environmental health (Creative City Network of Canada, 2005a, 2006b, 2007a). It is therefore essential to integrate cultural infrastructure into community building and broader planning and management processes, which means supporting cultural centers and facilitating cultural encounters (Creative City Network of Canada, 2005b). As Jon Hawkes (2001) contends: "No matter how commendable the values of a society may be, they amount to nothing if the society lacks life, vitality, dynamism and democratic public discourse...we need a process of nurture and cultivation" (p. 22).

Fariba Gharai (1998) observes that contemporary communication technologies such as Internet, webcam, and iPhone, to name a few, have enabled human interaction previously in a neighborhood extend to a global interaction with no distance limitation. However, the need for face-to-face interaction is still an important factor in the design of a neighborhood to enhance social relationships.

Duxbury and Jeannotte (2011) argue that cities and communities are socially and culturally constructed physical places within which patterns of life play out, and innovations and creativity are enabled by these spaces. Artistic and creative processes play important parts in social transition to a more sustainable future: they may enable new thoughts and insights, increase awareness and knowledge, and ultimately contribute to the transformation of individuals and communities (p. 6).

Hargreaves and Webster (2000) observe that activities in outdoor spaces between buildings are considered very important for forming social relations, and are seen as a necessary basis for building a community. It is the arrangement of spatial elements, buildings, paths, and green areas that provides the opportunities within which social interaction occurs (p. 6).

Courtyard houses have outdoor spaces that could be conducive to social relations (Broudehoux, 1994; Ekblad and Werne, 1990a; Ekblad *et al.*, 1992; Han, 2001; Kanazawa and Che, 2002; Liu, 1992; Lü, 1993; Sit, 1999; Tan, 1994, 1997; Wu, 1999; Zheng, 1995). This will be investigated further in the field research.

Cultural Vitality

Cultural vitality is the fourth and relatively new pillar in the sustainable development framework that refers to human wellbeing, creativity, diversity, and innovation. Local heritage and distinctiveness play a major role in facilitating cultural vitality because meaning and symbolism and its landscape contribute to the health and happiness of the people (Hargreaves and Webster, 2000; Hawkes, 2001).

Writers such as Darlow (1996) and Wheelwright (2000) observe that sustainable development is largely a cultural task since it seeks a change in attitudes and lifestyles. Judy Spokes, the executive officer of Cultural Development Networks, asserts that "Culture is both overarching and underpinning" (Hawkes, 2001, p. 3). As such, David Brand (2005, pp. 76-81) and Keith Nurse (2006, pp. 36-38) argue that culture should be placed at the front and center of the sustainability framework and fully incorporated into the other three pillars because it is a basis for questioning the implication and practice of sustainable development at its heart. Creative City Network of Canada (2005a) likewise contends that "Culture is a core dimension of vibrant and sustainable communities" (p. 1) because the character of a place is inseparable from its traditions and culture as they are lived and expressed in the activities and social life of the community; this quality of a city is one of its most salient features for making it a desirable place to live, work, study or visit.

Definition of Culture

Culture is one of the most complex English words because it may mean many different things to different people. It is commonly acknowledged that British anthropologist Edward B. Tylor (1832-1917) was the first Western scholar to advance a definition of culture in 1871. Tylor defined culture as "that complex whole which includes knowledge, belief, art, morals, law, custom, and any other capabilities and habits acquired by [hu]man as a member of society" (1871/1920, p. 1).

A more holistic definition of culture is found in *The Silent Language* (1959) by American anthropologist, Edward T. Hall (1914-2009), in which he explains, "culture has long stood for the way of life of a people, for the sum of their learned behavior, patterns, attitudes, and material things" (p. 31), and that culture "is not an exotic notion studied by a select group of anthropologists in the South Seas. It is a mold in which we are all cast, and it controls our daily lives in many unsuspected ways" (p. 38). This insight suggests that culture penetrates every part of human life.

The UNESCO's Universal Declaration on Cultural Diversity (2002) states that "culture should be regarded as the set of distinctive spiritual, material, intellectual, and emotional features of society or a social group, and that it encompasses, in addition to art and literature, lifestyles, ways of living together, value systems, traditions and beliefs" (UNESCO, 2002, p. 2; UNESCO World Report, 2009, p. 3; UNESCO-UIS, 2009, p. 9).

In the areas of anthropology, sociology, political science, and cultural geography, culture was initially referred to skilled human activities through which nonhuman nature is transformed. Culture is linked to words such as 'cultivate' and 'cultivation,' 'agriculture,' 'horticulture,' 'viticulture,' and 'apiculture,' all of which share a sense of growth, with its meaning metaphorically applied to human development. Enlightenment belief in human progress prompted attribution of such culture to the cultivation of body, mind, and spirit (Barnard and Spencer, 1997; Hawkes, 2001; Johnston *et al.*, 2000).

Among cultural critics, anthropologists, and sociologists, there are generally two interrelated definitions of culture. First, culture is the 'way of life' of a particular set of humans, their shared beliefs, values, ideology, language, knowledge, history, philosophy and religion, morals, norms, regulations of behavior, codes of manners, faith, traditions, conventions, customs, rituals, the social production and transmission of identities, meanings, aspirations, memories, purposes, attitudes, and understanding. It is a collective term for the symbolic and socially learned patterns, rather than biologically transmitted traits, by which human behavior can be distinguished from that of other primates (Abercrombie, Hill, and Turner, 2000; Bruce and Yearley, 2006; Hawkes, 2001; Jackson, 1989/1992; Jary and Jary, 2005; Johnson, 2000; Marshall, 1994, 1998; Outhwaite and Bottomore, 1993; Rapoport, 1969; Scruton; 1996).

Second, culture embraces the repertoire of human excellence and individual perfection exemplified by artistic and intellectual practices and product, such as dance, opera, music, painting, sculpture, furniture, architecture, landscaping or gardening, literature, poetry, dress, cuisine, science, technology, and other pursuits (Abercrombie *et al.*, 2000; Bruce and Yearley, 2006; Hawkes, 2001; Jackson, 1989/1992; Johnson, 2000; Johnston *et al.*, 2000; Outhwaite and Bottomore, 1993).

This study encompasses both aspects of the definition. To search for a balanced view of the term, it is worth knowing how the Chinese define culture.

Chinese Definition of Culture

The Chinese word for culture is 文化 (pronounced *wen hua*) existed in the ancient Chinese language. 文 refers to various colored, interlocking veins, and 化 implies the change and generation of form or the transformation of the nature of things that requires the improvement of behavior towards the better. The combination of 文化 first appeared in *Yi Jing* (The Book of Changes), stating that nature created male and female who married to form families, countries, and the world (Chinese Academy of Social Sciences, 2003). The Chinese also extend their kinship to include the natural world, with the metaphor that the sun is the father, the moon the mother, and the myriad things as their siblings first appearing in *Yi Jing*,[1] and later elaborated by Neo-Confucian philosopher Zhang Zai (1020-1078) in *Western Inscription* (西铭 quoted in the beginning of Chapter 1). The concepts of dualism, polarity, opposition, conflict, as well as reconciliation of contraries, were direct results of this solar and lunar symbolism. There seems to be no sharp distinction but an interdependent relationship between nature and culture in traditional Chinese culture.

Thus, the Chinese embrace two interconnected aspects of culture, 人文 (*ren wen*) and 天文 (*tian wen*). 天文 refers to the laws of nature (Dao), whereas 人文 refers to human conduct in society. An administrator of a country must observe the laws of nature to know the best time for farming, fishing, and hunting. They must also establish social order by maintaining correct hierarchical relationships between ruler-ministers, father-son, husband-wife, elder-younger siblings, and friends; conforming their behaviors to appropriate morals and rituals; and

---

1   乾, 天也, 故称乎父; 坤, 地也, 故称乎母.

extending them to the larger world beyond to establish a greater civilization[2] (Chinese Academy of Social Sciences, 2003).

In the Modern Chinese Dictionary, 文化 is defined as "the sum of material and spiritual wealth generated during the development of human society throughout history, it refers particularly to spiritual wealth" (my translation). This modern Chinese definition of culture somewhat echoes the Western notion mentioned earlier.

Definition of Cultural Sustainability

The cultural dimension of sustainability is less well defined and remains under-researched not only because culture has traditionally been seen as a component of the social dimension of sustainability, but also because it is difficult to measure. Culture is gradually emerging out of the domain of social sustainability and is being recognized as having a discrete, distinct, and integral part in sustainable development (Creative City Network of Canada, 2007b; Cultural Development Network, 2005; O'Shea, 2011). No matter whether culture is viewed as a fourth pillar alongside the other three, or as a central pillar, the expansion from the triple to the quadruple dimension characterizes the current sustainability debate.

The Sustainable Development Research Institute (1998) defines cultural sustainability as "the ability to retain cultural identity, and to allow change to be guided in ways that are consistent with the cultural values of a people" (p. 1).

A Canadian definition is that cultural sustainability is "the highest attainable level of creative expression and participation in cultural life, measured against the lowest impact/disruption to the environment, to social aspects of society, and to the economy" (Laroche, 2005, p. 43).

Other definitions of cultural sustainability are found elsewhere. Al-Hagla (2005) contends that "Cultural sustainability is mainly concerned with the continuity of cultural values linking all of the past, the present and the future" (p. 4). Spaling and Dekker (1996, p. 9) alternatively suggest that cultural holism should be a principle for cultural sustainability because cultures are holistic with internal integrity; they are systems made up of subsystems: political, economic, linguistic, religious, and so on, which interact with one another so that any change in one subsystem propels a shift in the others, and thus alters the system as a whole.

The UNESCO's Convention for the Safeguarding of the Intangible Cultural Heritage (2009) states that:

> 'intangible cultural heritage' is the practices, representations, expressions, knowledge, skills – as well as the instruments, objects, artefacts and cultural spaces associated therewith – that communities, groups and, in some cases, individuals recognize as part of their cultural heritage. This intangible cultural heritage, transmitted from generation to generation, is constantly recreated by

---

2   《易经·贲卦·象传》曰:"刚柔交错,天文也;文明以止,人文也。观乎天文以察时变,观乎人文以化成天下"。

communities and groups in response to their environment, their interaction with nature and their history, and provides them with a sense of identity and continuity, thus promoting respect for cultural diversity and human creativity. (p. 11)

'Safeguarding' means measures aimed at ensuring the viability of the intangible cultural heritage, including the identification, documentation, research, preservation, protection, promotion, enhancement, transmission, particularly through formal and non-formal education, as well as the revitalization of the various aspects of such heritage. (p. 12)

The 'intangible cultural heritage' defined by UNESCO (2009) is manifested in the following areas:

1. oral traditions and expressions, including language as a vehicle of the intangible cultural heritage;
2. performing arts;
3. social practices, rituals and festive events;
4. knowledge and practices concerning nature and the universe;
5. traditional craftsmanship (pp. 11-12).

UNESCO (2009, p. 6) links intangible cultural heritage with biological diversity and holds that sharing and disseminating this heritage encourages cultural exchange and understanding between peoples, constituting an asset for harmony and peace among human beings. The UNESCO's Decade of Education for Sustainable Development (2005-2014) further endorses cultural development as a tool for social policy to foster social equity, cultural diversity, public housing, health, and ecological preservation (Canadian Commission for UNESCO, n.d.; Duxbury and Gillette, 2007; Duxbury and Jeannotte, 2011). Creative City Network of Canada (2006a) recommends that one approach to cultural sustainability is to view it as an integrated planning and decision-making process that has a long-term perspective of the cultural development, vibrancy, and cohesion of communities.

In this study, cultural sustainability will be defined as the adaptation and transmission of the beneficial parts in a nation's material (tangible) and immaterial/ spiritual (intangible) culture that are conducive to the development of their present and future generations.

Cultural Diversity

Cultural diversity is as essential to cultural sustainability as biodiversity is to ecological sustainability (Hawkes, 2001; Merchant, 2004; UN World Commission on Culture and Development, 1995). Jon Hawkes (2001) contends that "Diverse values should not be respected just because we are tolerant folk, but because we must have a pool of diverse perspectives in order to survive, to adapt to changing conditions, to embrace the future" (p. 14). The UNESCO's Universal Declaration on

Cultural Diversity (2002) likewise asserts that "cultural diversity is as necessary for humankind as biodiversity is for nature" (article 1) because "creation draws on the roots of cultural tradition, but flourishes in contact with other cultures" (article 7).

Cultural diversity often provides a variety of critical worldviews to resolving problems in another culture than our own (Spaling and Dekker, 1996). For example, the knowledge of traditional medicine and human-environment relationships in non-western cultures has challenged western medical practice and environmental preservation. In *The Nature of Design* (2002), David Orr argues that "the only knowledge we've ever been able to count on for consistently good effect over the long run is knowledge that has been acquired slowly through cultural maturation" (pp. 38-39), and that "only over generations through a process of trial and error can knowledge eventually congeal into cultural wisdom about the art of living well within the resources, assets, and limits of a place" (p. 48).

In her final book, *Dark Age Ahead* (2004), Jane Jacobs asserts that the true power of a successful culture resides in its examples. To take this patient and mature attitude successfully, a society must be self-aware of its cultural wisdom. Otherwise, any culture that abandons the values that have given it competence, adaptability, and identity becomes weak and hollow (p. 176).

Richard Engelhardt, UNESCO Regional Advisor for Culture in Asia and the Pacific, similarly indicates:

> If a nation, a city, a community, or even a family is so unfortunate as to lose touch with its past – if we forget where we have come from, if we lose the map through time which history has drawn – if we lose our heritage, it will be impossible for us to chart where we are headed in the future. (UNESCO World Bank, 2000, p. 22)

Thus, cultural sustainability requires constructing the present and the future by adopting and absorbing past wisdom. The survival of local values should constitute the backbone of sustainability, especially in countries with traditional cultures (Özcan, Gültekin, and Dündar, 1998).

Indigenous cultures, cultural pluralism or multiculturalism, therefore should be emphasized because the diversity of languages, ideologies, literature, poetry, theater, films, music, arts, creativity, architecture, heritage conservation, festivals, and so on, are resources for cultural sustainability. Multiculturalism is more successfully implemented in some immigration countries such as Australia, Canada, and New Zealand than others such as the USA. Opponents of multiculturalism, for example, Samuel P. Huntington (1996/2011) in the USA, fear that such a concept may erode the country's social cohesion and may result in ethnic strife (pp. 305-307).

Nevertheless, Agenda 21 for Culture (2004) endorses multiculturalism because cultural diversity is the common heritage of humanity and a means to achieve a more satisfactory intellectual, emotional, moral, and spiritual existence (article 1). Moreover, Agenda 21 for Culture (2004) suggests ways to achieve cultural diversity, including for example, accessing the cultural and symbolic universe at all stages of life (article 13). Courtyard housing, which directly links Heaven and Earth, may help attain this goal.

A conference hosted by the Cultural Development Network in November 2004 in Melbourne, Australia, has produced a report *The Fourth Pillar of Sustainability: Culture, Engagement and Sustainable Communities* (2005) that showcases the impacts of multiculturalism on cultural diversity in three English-speaking countries receiving immigrants from all over the world: Australia, Canada, and New Zealand. Another conference that took place in December 2004, in Jerusalem, has generated a book *Cultural Education – Cultural Sustainability* edited by Bekerman and Kopelowitz (2008) that explores ways to preserve indigenous cultures by minority, diasporas, indigenous, and ethno-religious groups in multicultural societies through formal, non-formal, and informal education.

Some researchers (Duxbury, 2003; Simons and Dang, 2006; Yue, Khan, and Brook, 2011) have attempted to outline a set of cultural indicators as a measurement of cultural sustainability. However, Duxbury, Simons, and Warfield (2006) caution "against using these patterns as universally applicable factors" (p. 34) because cultural benchmarks vary from country to country, and from region to region.

Keith Nurse (2006) observes, however, that not all cultures are equal: some are more equal than others depending on the political and historical milieu. At one end of the pendulum, western technology is viewed either as the cause of or solution to the problem. At the other end, traditional, indigenous, especially non-western knowledge is either perceived as 'backward' and problematic, or romanticized as 'sacred wisdom' valued for future generations. There is a need to redress the global imbalance in the cultural sphere (p. 36).

The existence of cultural differences has often led to judgments about other societies based on one's own traditional standards. One has a tendency to view one's own conventions as 'right' and others as 'wrong.' An unspoken superiority has been assigned to Western culture because of its advanced technological and economic developments in general, although these advances have often accompanied colonialism and the exploitation of non-renewable resources.

Understandably, cultures have a spiritual value and not merely an economic one. If cultural relativism suggests that no final judgement decides whether one culture is superior to any other, we can only appreciate and evaluate the behavior of others in the context of their own cultures (Geertz, 1973; LeVine, 1984). As Gudykunst and Kim (1984) argue: "all cultures are of equal value and their values and behaviors can only be judged using that culture as a frame of reference" (p. 97).

Spaling and Dekker (1996) contend that cultural change is not only inevitable, but also essential and desirable. There are aspects of all cultures that are either destructive or oppressive, resulting in disharmony among individuals, communities, and the cosmos. It has been recognized that cultural values need to be changed when they violate the integrity of people, groups, or the creation.

Cultural Activities

Cultural activity is an individual or group activity that involves promoting participants' values and belief systems. Cultural activity can likewise be an individual or group activity pursued for individual excellence, artistic or intellectual practices and product. In the Chinese context, cultural activities may include maintaining health/natural healing (*taiji*, *qigong*, martial arts, etc.), gardening, landscape painting, practicing calligraphy, composing poetry/essay, drinking tea, playing games (chess/checker, *weiqi*/Go game, *majiang*, etc.), playing traditional musical instruments, singing Chinese opera, celebrating birthdays, holding wedding ceremonies, and so on. As Creative City Network of Canada (2006a) states: "Cultural activities are recognized as an important way for individuals to contribute to their community.... Wide, inclusive participation in cultural activities contributes to community vitality and supports the four-pillar model of sustainability" (p. 1).

UNESCO-UIS (2009) also announces that "Cultural activities embody or convey cultural expressions, irrespective of the commercial value they may have. Cultural activities may be an end in themselves or they may contribute to the production of cultural goods and services" (pp. 23, 87). Jon Hawkes (2001) likewise maintains that "active community participation in arts is an essential component of a healthy and sustainable society" (p. 2) and that "a healthy society depends, first and foremost, on open, lively and influential cultural activity amongst the communities within it; sustainability can only be achieved when it becomes an enthusiastically embraced part of our culture" (p. 25). Thus, cultural activities function as a catalyst and indicator for cultural sustainability.

Courtyard housing has traditionally facilitated lively cultural activities, festivities, and rituals associated with birth, marriage, and death. One way of holding wedding ceremonies is to be in the courtyard so that marriage could be witnessed by heavenly bodies (Noble, 2003; Randhawa, 1999; Sinha, 1994). Celebrations of the seasons have been important festivities to take place in courtyards because a courtyard offers direct observation of the cyclic movements of sunlight, the waxing and waning of the moon, as well as an immediate connection with nature (Chapter 5). These practices can still be observed in many courtyard houses in China and elsewhere (Giedion, 1981; Ujam, 2006).

Political Support

A growing body of literature suggests that governance, policy, and political agenda are crucial factors in redirecting us towards environmental, economic, social, and cultural sustainability (Alred, 2008, p. 3; Brand, 2005, pp. 76-81; Cultural Development Network, 2005, p. 4; Duxbury and Jeannotte, 2011; Duxbury, Simons, and Warfield, 2006; Hawkes, 2001, p. 16; Holt, 2005, p. 40; Illa and Ulldemolins, 2011; Kingma, 2005, p. 63; Laaksonen, 2006; Laroche, 2005, p. 42;

Mercer, 2006; O'Shea, 2011; Savova, 2011; Stanborough, 2011; United Cities and Local Governments, 2010, p. 2; UNESCO, 2006; Ursic, 2011). However, the political dimension of sustainability has not but should have been warranted as a fifth pillar of sustainable development.

Culture is undoubtedly linked to and affected by polity. From the past to the present, China's political authority has always played a fundamental role in determining relations between human society and the natural world (Van Elzen, 2010, pp. 9, 82; Weller and Bol, 1998, pp. 313, 328, 336-337). The slow pace of social and technological development in China before 1978 has likely been related to a political structure that still carried traces of feudal ideologies and methodologies. The May Fourth Movement of 1919 (often referred to as the New Culture Movement, 1915-1921) was primarily political, but it was also cultural because it called for an end to the patriarchal family in favor of individual freedom and women's liberation. Through a re-examination of Confucian texts and ancient classics using modern textual and critical methods (known as the Doubting Antiquity School), it established democratic and egalitarian values, and launched an orientation to the future rather than the past, to name some initiatives among many.

Since the 1920s, China has increasingly continued the destruction of the old world and traditional culture (Van Elzen, 2010, pp. 94, 167-168). A speech by Mao Zedong on Literature and Art (in 1942) claimed that "In the world today all culture, all literature and art belong to definite classes and are geared to definite political lines" (translated by Schram, 1967, p. 299). Hence, the Cultural Revolution (1966-1976) was only 'cultural' in name; in reality it was a power struggle involving Mao (see, for example, Burns, 1978, pp. 402-403; Michael, 1977; Mohanty, 1978; Schram, 1971; Schrift, 2001; Terrill, 1999; Uhalley, 1975; Van Elzen, 2010, p. 83; Wilson, 1979). Mao carried forward his efforts to demolish traditional China and replace it with his own Maoist system that eventually imploded. When Maoism collapsed, the Chinese people went back in search of a moral framework in traditional culture; their pursuit may explain an increasing interest in Confucianism, Daoism, Buddhism, and Christianity in China today (Van Elzen, 2010, pp. 94, 167-168). However, the lasting consequences of the Cultural Revolution point to a total breakdown of traditional Chinese cultural patterns and education (Gao, 1999; Van Elzen, 2010, pp. 50, 80-81; Weller and Bol, 1998, p. 325).

The Cultural Revolution also had a significant impact on traditional Chinese dwelling culture, because the Chinese Communist Party also disregarded *Feng Shui* theory (Chapter 3) which has influenced imperial Chinese city planning and classical courtyard house design for centuries. Nonetheless, there has been a growing interest in *Feng Shui* in recent years, which can be seen from the numerous publications of books being sold in bookstores in China and elsewhere and the number of research papers published in journals (Knapp, 2005c; Weller and Bol, 1998, p. 332).

Ch'ü (1994) reports an extensive survey conducted in Taiwan in 1994 that about 50 percent of the population considered *Feng Shui* important for heath, wealth, and success of their children's future, and that about 80 percent indicated that

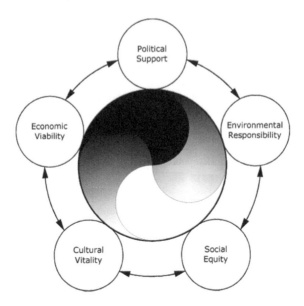

Figure 2.3   A conceptual sustainability framework inspired by *wuxing* (Five Phases). The core is my expanded *Yin Yang Yuan* symbol (explained in Chapter 3) representing the cyclic movements of sunlight and the four seasons. Drawing by Donia Zhang 2011

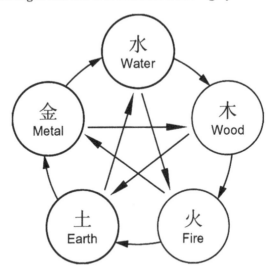

Figure 2.4   Wuxing (Five Phases) in Chinese philosophy describing the mutual generative order (outer arrows) and mutual destructive order (inner arrows) of the five natural elements. Drawing by Donia Zhang 2009

they would consult a *Feng Shui* expert to choose appropriate times for weddings, opening a new business, or moving to a new house (pp. 165-166). In recent decades, *Feng Shui* has gained a global popularity in the design of housing, offices, and medical buildings in China, Europe, and America (Knapp, 2005a, 2005d; Orr, 2002; Rapoport, 2005). *Feng Shui* is a harmonious ethic with a scientific basis that may be helpful in addressing the environmental crisis (Anderson, 2001, p. 172; Knapp, 1999, p. 167).

With a current revival of Chinese philosophy, the Confucian *Analects,* along with other traditional Chinese thought, have been taught on China Central Television and some private schools in China (Krieger and Trauzettel, 1991; Tucker and Berthrong, 1998). Confucius has been featured in new books, and a 9.5 m bronze statue of the scholar was erected in 2010, northeast of Beijing's Museum of Chinese History, and opposite Tiananmen Square, approximately facing the conspicuous portrait of Mao Zedong. However, the sculpture was mysteriously relocated three months later to the enclosed garden of the Museum (Bodeen, 2011). This action implies disagreement at the higher, administrative level about reviving Confucius as both an underlying ideology and a renewed cultural icon in China.

Thus, the environmental, economic, social, cultural, and political dimensions of sustainability are closely intertwined; without political support, it is simply impossible to achieve cultural sustainability.

Summary and Conclusion

This chapter is a background study on cultural sustainability from a literature review. It takes the stance that cultural vitality is the fourth pillar of sustainable development alongside the other three: environmental responsibility, economic viability, and social equity. The chapter also argues that political support should be regarded as an independent, fifth pillar because government as the superstructure makes vital decisions; without political support, cultural sustainability is simply impossible to achieve. Nevertheless, the study focuses on cultural sustainability.

In Chapter 3 that follows, the aspects of Chinese philosophy that have influenced imperial Chinese capital city planning and classical courtyard house design will be discussed, to set the context of the study and draw cultural principles as benchmarks against which the renewed/new courtyard housing projects will be measured in Chapters 7-10.

Chapter 3
Four Key Themes in Chinese Philosophy on City Planning and Courtyard House Design

He who has a clear understanding of the Virtue of
Heaven and Earth may be called the Great Source, the Great Ancestor.
He harmonizes with Heaven;
and by doing so he brings equitable accord to the world
and harmonizes with men as well.
To harmonize with men is called human joy;
to harmonize with Heaven is called Heavenly joy.

– Zhuangzi, c.369-286 BCE, *The Complete Works*, Chapter 13

This chapter identifies four key themes in Chinese philosophy that have influenced city planning and classical courtyard house design, with a view to using these principles as benchmarks for measuring the renewed/new courtyard housing projects examined in Chapters 7-10, to see if they are culturally sustainable. The chapter consists of four sections: harmony with heaven, harmony with earth, harmony with humans, and harmony with self.

In Chinese philosophy, *Yin Yang* balance and harmony is a fundamental concept applied to both nature and human affairs. *Yin Yang* (阴阳) literally means 'shade and light' with the word *Yin* (阴) derived from the word for 'moon' (月) and *Yang* (阳) for 'sun' (日). *Zhou Yi* (*Yi Jing*, Book of Changes) suggests that polar opposites created Heaven and Earth, *Yin* and *Yang*. When Heaven and Earth intersect and *Yin* and *Yang* unite, it gives life to all things. When *Yin* and *Yang* separate, all things perish. When *Yin* and *Yang* are in disorder, all things change. When *Yin* and *Yang* are in balance, all things are constant.[1] The mutual interdependence of *Yin* and *Yang* is called 和合 (*hehe*). The first 和 signifies 'harmony' or 'peace,' and the second 合 denotes 'union' or 'enclosure.' The combined words imply that harmonious union of *Yin* and *Yang* will result in good fortune, and that any conflict is viewed only as a means to eventual harmony (Lau, 1991, p. 214; Yu, 1991, pp. 51-52).

---

1   阴阳合则生, 阴阳离则灭; 阴阳错则变, 阴阳平则恒.

Winter Solstice

Autumn Equinox 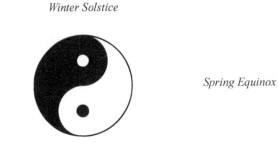 *Spring Equinox*

Summer Solstice

Figure 3.1   The most popular *Yin Yang* symbol. It is said the symbol was revised by Daoist sage Chen Tuan (陈抟, 872-989 CE) based on the *taiji* image in *Yi Jing*

Yin Yang symbol is a Chinese representation of the celestial and terrestrial phenomena, it is a schematic map to illustrate the Dao (Way) and imply completeness, represented by the shape of a circle. The two fish embracing one another signifies their mutual generation and interdependence. The two dots symbolize that each time one of the two forces reaches its extreme, it contains in itself the seed of its opposite (Capra, 1975/1999, p. 107). However, the symbol as a representation of universal phenomena is limited to duality.

Harmony with Heaven

Harmony with Heaven received primary attention in Chinese philosophy, with roots that can be traced back to *Yi Jing* (The Book of Changes), that first flourished around 6th century BCE, and that draws on an oral tradition in the Neolithic period (beginning c.9500 BCE) (Yang, 2008). *Yi Jing* concerns human conduct in accordance with the universe to bring good fortune, and is the most influential book in Chinese culture. It is commonly held that *Yi Jing* was written by the legendary Chinese emperor *Fu Xi* (2953-2838 BCE), with further commentaries added by King Wen and the Duke of Zhou in the 11th century BCE, and that Confucius (551-479 BCE) later revised it (Yang, 2008).

Yi Jing establishes a mathematical structure that depicts eight natural phenomena: Heaven (乾 *qián*/天 *tiān*),[2] Earth (坤 *kūn*/地 *dì*), thunder (震 *zhèn*/雷 *léi*), water (坎 *kǎn*/水 *shuǐ*), mountain (艮 *gèn*/山 *shān*), wind (巽 *xùn*/风 *fēng*), fire (离 *lí*/火 *huǒ*), and marsh (兑 *duì*/泽 *zé*). It uses *Yin Yang* strokes ("–"for *Yang* and "– –" for *Yin*) to explain myriad things and their combination generates eight trigrams (*bagua*) and 64 hexagrams. Their interaction, transformation, growth,

---

2   The first character is the classical Chinese term and the second is the modern interpretation.

and decline rationalize how things take their forms and change with time (Ma, 1999).

Supported by *Yi Jing* and founded on observations of nature, the ancient Chinese developed *Feng Shui* ('Wind and Water') theory, that dates back 5,000 years ago, for selecting appropriate sites for cities, villages, houses, and cemeteries. Because of *Feng Shui*, the people's attentiveness to environmental conditions was significantly heightened (Knapp, 2005a, 2005d; Orr, 2002; Rapoport, 2005; Walters, 1991).

To achieve the 'unity of heaven and humans' (天人合一), *Feng Shui*'s basic principle is to establish an optimum *Yin Yang* balance when selecting a site by matching 'the right time, the right place, and the right people' (天时, 地利, 人和) and by observing the surrounding environments carefully and thoroughly, to be in harmony with nature and to modify and utilize nature when creating favorable conditions for human habitat.

Nine Constellations Magic Square Matrix

The ancient Chinese considered that Heaven was a circle (*Yang*) in shape and the Earth a square (*Yin*). Hence, in their design of cities and buildings, they preferred combining circles (dynamic) with squares (static) to establish balance and harmony, and avoided sharp figures such as triangles or diamonds that point directly to the sky (Kou, 2005; Yang, 2008). The Chinese also believed that Heaven had nine fields, Earth had nine continents, land had nine mountains, and mountains had nine paths. Thus the number 9 was a supreme figure, and the Earth was represented by a large square divided 3×3 into nine smaller ones.

The cosmology of classical Chinese courtyard houses was gradually established in the Han (206 BCE-220 CE) and Tang (618-907) times, when the Chinese believed that they could achieve harmony, stability, and good fortune by residing in an environment following the principles of cosmic order. The correct orientation was essential to houses because the Chinese worshipped the Sun and Moon gods. They took three main steps towards maintaining cosmic order. First, find a center of Earth that was connected to the center of Heaven by a cosmic axis. Second, locate east and west and correctly position the worship platforms for the Sun and Moon gods, and find the north-south axis from the position of the North Star. Third, orient the house according to the four cardinal directions. The central courtyard provided an opening to the sky, allowing the drift of smoke to heavenly gods, sunlight, and a pit in the ground for worshipping earth gods. This cosmological thinking dominated the arrangement of most Chinese houses until the end of the traditional period (1911); however, it limited the possibilities of creating more variety of house forms (Chang, 1986, pp. 167-169, 204).

North

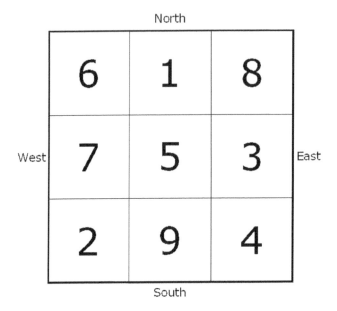

West   East

South

Figure 3.2   The Nine Constellations Magic Square Matrix (九宫图) that influenced Chinese city planning and classical courtyard house design. Drawing by Donia Zhang 2009

Sunlight and Situating Buildings

Orientation is an important factor in classical Chinese courtyard house design because of the quality of sunlight which can provide many benefits to humans:

1. In the northern hemisphere, sunlight can warm people in winter, and the temperature of south-facing rooms is higher than north-facing ones (in the southern hemisphere, north and south reverses);
2. Sunlight participates in the synthesis of Vitamin D in human bodies that can prevent rickets in children and help elderly people strengthen their bones;
3. A moderate amount of ultraviolet rays in sunlight can kill bacteria, and especially good for people with respiratory diseases; and
4. Sunlight can improve human immunity (Kou, 2005; Luò, 2006).

This is why *Feng Shui* theory advocates that houses should 'sit north and face south,' 'carry *Yin* at the back and encompass *Yang* in the front,' or 'have hills behind and river in front,'[3] not only for receiving more sunlight, but also for avoiding chilly winds (Kou, 2005; Luò, 2006; Ma, 1999).

---

3   负阴抱阳, 背山面水.

Wind directions in China vary with the season, cold currents come from Siberia in the winter and breezes blow from the Pacific Ocean in the summer. In *Feng Shui* theory, wind is differentiated as *Yin* or *Yang* wind. Warm east and south winds are *Yang* and harmless, whereas cold west and north winds are *Yin* that should be avoided or they could chill to the bone; only soft winds are desirable (Kou, 2005; Luò, 2006; Ma, 1999).

For urban sites without mountains or rivers, the surrounding roads can be regarded as waterways, the enclosing walls as mountains, and the roofs opposite the house as a mountain form. These extended meanings have made *Feng Shui* theory more widely applicable (Ma, 1999, p. 31).

Feng Shui masters employ a compass as their primary device for finding auspicious orientations. The compass combines time and space and divides the universe into heaven, earth, and humans, among which it sets up an analogical relationship to determine the intrinsic connection of the three elements and offers suggestions for various circumstances. Due to limitations of scientific development, it is inevitable that the readings from the compass may lead to subjective views (Knapp, 2005a, 2005d; Kou, 2005; Luó, 2006; Luò, 2006; Ma, 1999; Pan, 2002). To my knowledge, a compass is still used by some people in China and Canada when choosing an auspicious house.

Qi and Water

Qi ('matter-energy' or 'material force') is the core concept in *Feng Shui* theory and is considered as a primary substance that all things are made of, including humans. Nowadays, *qi* is perceived as the microwave radiation and celestial electromagnetic radiation from the sun (Kou, 2005; Luò, 2006). *Qi* is thought to be constantly changing to become mountains and water, and is moving above the air and below the earth to nourish all things (Luò, 2006; Ma, 1999).

The Chinese have long recognized there is a closed circulatory system among groundwater, wind, cloud, and rain as groundwater evaporates into winds, winds rise up to become cloud, cloud turns into rain, rain penetrates the ground to become groundwater again, and water collects cosmic *qi*. In selecting a site, *Feng Shui* masters pay particular attention to the relationship between water and *qi* and contend that a site surrounded by mountains and encircled by water must have *qi* because water absorbs microwaves easily, and when *qi* meets water, it creates an enclosure conducive to human health (Luò, 2006; Ma, 1999). Thus, water is a favorable feature in a city or near a house.

Exterior Form

Feng Shui advocates an enclosed space because it can 'hide winds and gather *qi*' (藏风聚气). The symmetrical and complete form of classical Chinese courtyard houses psychologically fulfilled human desire for perfection (Ma, 1999; Xu, 1998). *Feng Shui* theory considers a house's exterior environment as the 'tree trunk' and its interior as the 'branches,' and this order should not be reversed because the *Ten*

Books on Houses for the Living (阳宅十书) states: "If the overall form of a house is inauspicious, no matter how properly arranged the interior spaces, it is generally inauspicious. Thus the exterior form of a building should be of primary concern"[4] (Ma, 1999, p. 31; my translation).

If a house complex is longer in the south-north direction with a high and wide northern hall, it will obviously benefit from more sunlight and natural ventilation. Adjacency to water (or roads) brings convenience to life, such as through irrigation, transportation, and aquatics. A gently sloped site avoids flooding and makes drainage easier. The luxuriant vegetation on a mountain conserves water and soil, modifying the microclimate, and providing fuel and resources. This relatively enclosed natural environment may positively contribute to forming an ecologically good life conducive to the health and wellbeing of its occupants (Ma, 1999, p. 28).

The location of a house gate is important in *Feng Shui* theory since a gate is the 'mouth of *qi*' (气口) through which *qi* enters or exits. Its orientation and size of opening have direct links to communication between the inside and outside. If there is a waterway or road encircling the gate, it will gain *qi* and exchange information with the cosmos, which will contribute to the occupants' health. If a gate is in a wrong place, it will weaken *qi* and destroy the intimate atmosphere of the house. Classical Chinese courtyard houses situated on the northern side of east-west running lanes normally have gates in the southeast corner to greet the morning sun and to echo a Daoist motto 'purple *qi* comes from the east' (紫气东来). The houses situated on the southern side of lanes often have gates in the northwest corner (Kou, 2005, p. 91; Ma, 1999, p. 30).

Courtyards

If the exterior space enclosed by walls is moderate and the courtyard size proportional to the height of surrounding buildings, the occupants will feel comfortable. If it is out of proportion, the inhabitants will feel distant or constrained, and the microclimate will need adjustment by building higher or lower enclosing walls to improve air circulation in the courtyard (Ma, 1999, p. 30). Moreover, a single tree in the courtyard should be avoided because wood (木) inside a square (口) resembles the word 困 ('being held') (Kou, 2005, pp. 49, 81).

Harmony with Earth

> *Here, where Heaven and Earth are in perfect accord,*
> *where the four seasons come together,*
> *where the winds and the rains gather,*
> *where the forces of Yin and Yang are harmonized,*
> *one builds a royal capital.*
>
> — *Yi Jing*; cited in Wu, 1999, p. xi

---

4   若大形不善, 总内形得法, 终不全吉, 故论宅外形第一.

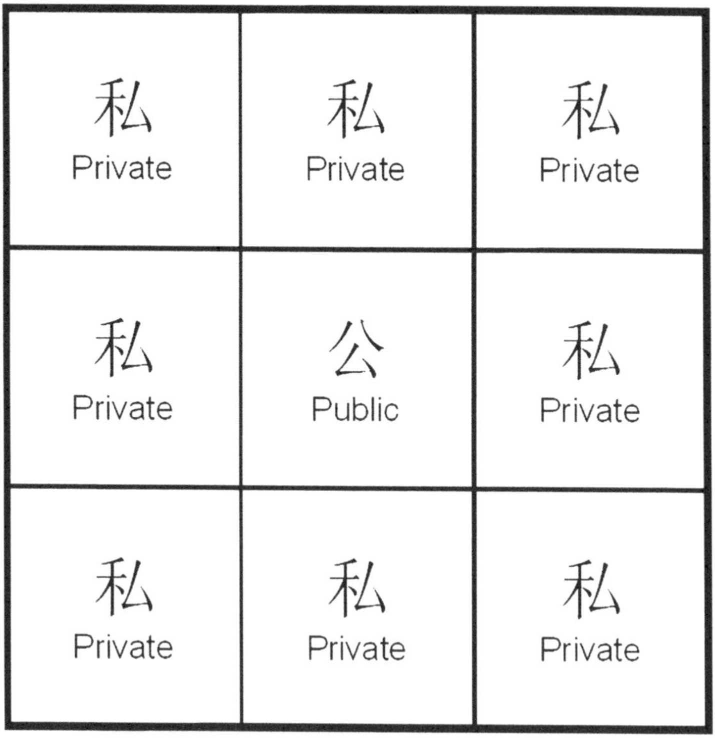

私 Private	私 Private	私 Private
私 Private	公 Public	私 Private
私 Private	私 Private	私 Private

Figure 3.3   The Nine Squares System (井田制). Drawing by Donia Zhang based on the *Complete Chinese Dictionary*, 1915; *Chinese-English Dictionary*, 1980

Harmony with Earth is another significant concept in Chinese philosophy as Daoists advise us not to dig or gouge Mother Earth because when her body is exploited, the earth *qi* (地气) leaks and consequently her suffering and resentment may bounce back to humans (Lai, 2001, pp. 104, 106). Earth *qi* refers to favorable size, level, soil quality, temperature, and humidity of the site which is utterly important to the occupants' physiology. If the earth *qi* is too strong or too weak, it may make the residents feel uncomfortable, which requires the adjustment of *qi* entering from the gate, door, or window (Luò, 2006).

Nine Squares System

Founded on the Nine Constellations Magic Square Matrix (九宫图; figure 3.2), the Nine Squares Land Ownership System (井田制; *Yi Jing*, Hexagram 48; figure 3.3) has guided imperial Chinese capital city planning and classical courtyard house design. The system is represented by the word 井 (*jing*), the symbol of a water well signifying a portion of land divided into nine parts, with the central square (630 *mu* or 42 ha) [5] belonging to the public and cultivated by the joint labor of the eight private property owners settled on the other divisions who shared the water well at the center. The system started during the Yellow Emperor's regime (2697-2598 BCE) (*Complete Chinese Dictionary*, 1915, p. 72; *Chinese-English Dictionary*, 1980, p. 362) and is still widely used in agricultural areas of China (Lok Tok and Zhang Junmin, personal communications, 2007).

Ideal Capital City

Historical records show that when the Chinese first invented the word for walled enclosures in the 3rd millennium BCE, they named it 城 (*cheng*), meaning 'city' nowadays (Schinz, 1989). An imperial Chinese capital city was always enclosed by walls, a symbol of strength and a means of defence against attackers.

All imperial Chinese capital cities had a central axis symbolizing the access to Heaven, along which were located the Imperial Palace, government buildings, Bell and Drum Towers (early timekeeping devices), and other public buildings. A passage from the *Record of Trades* in *Rituals of Zhou* (周礼•考工记, c.1066-221 BCE) says:

> The master craftsman constructs the state capital. He makes a square of nine *li* [3,735 m], where there are three gates on each side. Within the capital, there are nine north-south and nine east-west streets. The north-south streets have a width of nine-carriage tracks, with the Palace at the center, the Ancestral Temple on the left (east), and the Altars of Soil and Grain on the right (west). In the front (south), there is the Hall of Audience, and the markets behind (north). [6] (My translation)

An illustration from the *Record of Trades* in *Rituals of Zhou* (c.1066-221 BCE) shows this ideal capital city in imperial China (figure 3.4): it is divided into *Yin* and *Yang* sections with the imperial courts located in the auspicious southern half (*Yang*) and the markets in the less favorable northern half (*Yin*). [7] The Confucian *Doctrine of the Mean* suggests that when *Yin* and *Yang* are in balance, it is 'proper.' Properness also implies to live at the center. Thus the superior power and centrality of the emperor as the 'Son of Heaven' is manifested in the central location of the Imperial Palace, and all imperial Chinese capitals were located approximately at the center of their states, with the supreme number 9 associated

---

5   1 *mu* = 0.066667 hectares = 666.67 sqm

6   In China in the past, the drawings traditionally had south orientation upwards and north direction downwards, but it is no longer the norm in China nowadays.

7   In imperial China, markets were considered inferior spaces due to Confucian devaluing of money-making. The markets situated behind the Palace may be because the emperor at the center of the city would not have to face them.

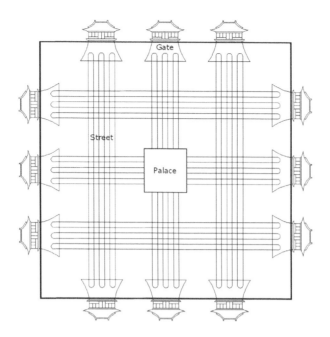

Figure 3.4   **The ideal capital city in imperial China from the *Record of Trades* in *Rituals of Zhou* (c.1066-221 BCE). All the measurements within it are multiples of heavenly number 3. Drawing by Donia Zhang after Liu, 2002a**

with the emperor (Beijing City Planning Committee, 2002; *History of Chinese Architecture*, 1986/2009; Kou, 2005, p. 172; Liu, 2002a; Pan, 2002; Qiao, 2002).

All imperial Chinese capital cities strove to achieve this idealized form but no city in China is found *exactly* matching this plan. The closest example is Qufu, the capital of Lu in today's Shandong province where Confucius was born and lived, and where his temple was built in 478 BCE (*History of Chinese Architecture*, 1986/2009; Pan, 2002, p. 226). The application of Chinese philosophy is well exemplified in the planning of the two case study cities: Beijing and Suzhou, as described below.

Beijing's City Planning

Beijing (literally means 'Northern Capital') as a city has a history of over 3,000 years. It was ruined and rebuilt in 1274 as the capital of the Yuan dynasty (1279-1368) and has remained the capital of China since then. Geographically, Beijing (latitude 39°55'N, longitude 116°25'E) is situated in the northeast of China with a 'continental monsoon' climate of four distinct seasons, its prevailing winds come from the southeast in the summer and the northwest in the winter.

Unlike naturally evolved European capital cities such as London, Beijing was initially a planned city applying principles described in the *Record of Trades* in *Rituals of Zhou*, with a 7.8 km north-south central axis[8] denoting the greater axis from Heaven to Earth, and the Purple Forbidden City, home of 24 emperors from the Ming (1368-1644) and Qing (1644-1911) dynasties, located at the center. The Palace City was enclosed on four sides by the Imperial City, which was further enclosed on four sides by the outer-city walls (Guo, 2002; Luó, 2006).

In Daoist legend, the celestial Palace of the Jade Emperor (紫微宫) is a star near the Polaris and was perceived to be the center of the universe. Hence, the Chinese emperors had their Purple Forbidden City built at the center of Beijing to reflect this celestial center. The color purple and radiance of the Polaris, symbolizing the center of the universe, heighten the sacredness of the Imperial Palace. The arch-shaped Golden Pond in front of the Gate of Heavenly Peace encircling the Forbidden City satisfies the desire of constantly incoming wealth (Luó, 2006). By maintaining the central axis and continuing to place buildings symmetrically along the axis, Beijing has preserved the image of imperial power (Sun, 2002).

Beijing was planned with a water dragon (*Yin*) that is winding, flowing, and active, and a land dragon (*Yang*) that is straight, ordered, and static, they complement each other. The land dragon is the Central Axis and the dragon head is the Gate of Heavenly Peace and its left foot the Ancestral Temple (now Cultural Palace), the right foot the Altars of Soil and Grain (now *Zhongshan* Park), and its body the Forbidden City, *Jingshan* Park (Coal Hill), and the Drum and Bell Towers. The water dragon is the Shichahai Lake District to the west of the land dragon and the dragon head is Nanhai ('South Lake'), its body from Zhonghai ('Central Lake') to Beihai ('North Lake'), Qianhai ('Front Lake'), Houhai ('Rear Lake'), and Xihai ('West Lake') (Luó, 2006).

Heaven and Earth were worshipped at Beijing suburban altars on winter and summer solstices between 1369 and 1375. The Temple of Heaven was first built in the south of Beijing in 1420, where sacrifices to Heaven and Earth were performed in the same place. However, it was thought that the combination of these two temples violated the *Rituals of Zhou*. Subsequently, in 1530, the Temple of Earth (north), the Temple of Sun (east), and the Temple of Moon (west) were erected to define the border of this imperial capital (Pan, 2002). The Beihai Park in inner Beijing and the Summer Palace in the western suburb are landscaped

---

8   Recent scholarly observations found that this central axis does not exactly line up with the south-north meridian line, but is around 2 degrees off to the west. There have been several explanations about it and the more convincing one is that it connects to the Yuan Shangdu (元上都) in inner Mongolia, where the Yuan dynasty ruler Kublai Khan (1215-1294) was enthroned and originally came from (Across China, May 9, 2010). Or maybe that was the direction of magnetic north at the time. It has been suggested that the slightly different alignment of parts of some ancient churches that were built at different times is related to the shifts of magnetic north (Michael Humphreys, personal communication, 2011).

imperial gardens with lakes imitating southern Chinese gardens, although their architectural grandeurs are typical of the northern type.

During the Liao time (907-1125), Beijing was divided into nine wards (坊 or 里坊). When the Jin Empire (1115-1234) expanded their capital, the streets and lanes cut through wards to loosen the four-sided enclosures that had defined the ward system, and a more informally arranged array of streets and roads formed 62 neighborhoods (坊巷). A north-south imperial thoroughfare divided the capital into east and west districts, resulting in an increase in commercial activities and a strong spirit of commercialism (Guo, 2002).

The destruction of the Jin Zhongdu ('Central Capital,' previous name of Beijing) in 1215 was a starting point for reconstructing a new capital in 1267 renamed Yuan Dadu ('Great Capital,' previous name of Beijing). The rebirth of a triple-walled capital signifying a new generation of Chinese imperial city followed the principles in the *Record of Trades* in *Rituals of Zhou*, with a stone laid marking the center spot of the city. The Palace and Imperial City were pushed southward from the center (figure 3.5), with the Bell and Drum Towers positioned along the central axis. The residential quarters occupied the northern half in two areas, each with 62 and 75 residential wards, respectively. Yuan Dadu was thus more strictly bordered by wards than its Song predecessors. Yuan Dadu was renamed Beiping ('Northern Peace') after the establishment of the Ming dynasty (1368-1644) and renamed Beijing ('Northern Capital') in 1403. Beijing's reconstruction was executed between 1370 and 1553, with the outer boundary extended and an inner concentric Palace City rebuilt (Pan, 2002). Much of Beijing's inner city today has continued the urban structure of the Ming and Qing dynasties (1644-1911) with no fundamental alteration in layout (Beijing City Planning Committee, 2002).

Beijing's inner city has a grid pattern like a chessboard. Until recently, inner Beijing was divided into four quarters: Eastern District, Western District, Chongwen District (merged with Eastern District in 2010), and Xuanwu District (merged with Western District in 2010).

Beijing's traditional courtyard houses (*siheyuan*) are embedded in its urban fabric (Wu, 1999; figure 3.6), and the wealthy lived in the Eastern District and the noble in the Western District (东富西贵) in the Qing dynasty (1644-1911), though the situation had changed in late Qing. The residential lanes, known as *hutong*, mostly run east-west, and streets north-south. The neighborhoods are grouped in two main sectors with a total of 62 sqkm. The more important of the two is the 38-sqkm inner city, originally enclosed by old city walls which were replaced by the second ring road in the 1960s (the first ring road refers to the walls of the Imperial City, of which only some fragments remain). The second historic area is the 24-sqkm outer city located south of Qian Men (Front Gate) that was originally a separate walled city.

Preeminent Chinese architects Liang Sicheng and Chen Zhanxiang in the 1950s proposed to build another city center alongside the old, to better preserve the historic city. However, the idea was discarded by former Chairman Mao Zedong who favored the Russian 'experts' plan to readjust and accommodate the

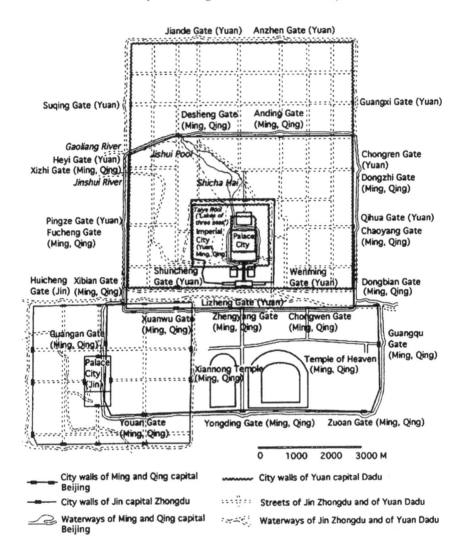

Figure 3.5   Evolution of Beijing since the 12th century with a chessboard urban pattern persistently remaining throughout the dynastic period: Jin Zhongdu ('Central Capital'), Yuan Dadu ('Great Capital'), and Ming and Qing Beijing ('Northern Capital'). Source: Wu, 1986

Figure 3.6   Beijing's traditional courtyard houses were prevalent, as shown
              in this photo taken probably in the 1980s. Source: Ma, 1993, p. 2

new government in the old city center. Parts of the inner-city walls and most of
the outer-city walls and gates were demolished in 1955-1958 to facilitate traffic
flow. The remaining inner-city walls and gates[9] were destroyed in 1965-1969 to
construct Beijing's subway (J. Wang, 2003, p. 313).

Liang and Chen's original vision of making Beijing's city walls 'the world's
unique three-dimensional round-the-city garden' (Liang, 1998, p. 307) was
unrealized because the ideology at the time regarded the old city as traces of feudal
societies that should be revolutionized, transformed, and firmly smashed. Thus,
this originally scholarly debate became political (J. Wang, 2003, p. 230). Liang
and Chen's scheme of building a new city center alongside the old was in fact
more economical as, even today, it is still less expensive to build a new district

9   Of the 16 pairs of city gates, only one and a half pairs remain: the *Zheng Yang Men*
gate tower and watchtower, and the *De Sheng Men* watchtower (J. Wang, 2003, p. 313).

than to renovate the old.[10] Beijing has since then expanded and it is now organized in a series of seven (there have been talks about eight and nine) ring roads.

On the discourse of 'Beijing Master Plan 2004-2020,' Professor Wu Liangyong has proposed a multi-city-center strategy to 'step out of the concentric circle' (Wu, 2005a, p. 25; 2005b, p. 70), and to alleviate overcrowding.

Suzhou's City Planning

Suzhou has a rich history and culture of 2,500 years. During the Spring and Autumn periods in 514 BCE, Suzhou was built as the Wu State (吴国) capital by Wu Zixu (伍子胥, ?-484 BCE) who "examined the soil and tasted the water, and mimicked Heaven and followed Earth"[11] (my translation). Suzhou's old city location has not shifted since its inception because of the wise choice of the site. Geographically, Suzhou (latitude 31°19' N, longitude 120° 37'E) is situated in the southeast of China on the shores of Lake Tai in the lower reaches of Yangzi River and enjoys a mild and humid subtropical climate with plum rains[12] in June and July (Orasia, 2002; Xu, 2000; Yu, 2007).

Suzhou's prosperity began after the construction of the Jinghang Grand Canal in the Sui dynasty (581-618) and the city was named Pingjiang ('Peace River') in the Song dynasty (960-1279). The Pingjiang map (dated 1229; figure 3.7) is the oldest and most complete ancient city plan in China, offering an important source for study the transformation of Chinese cities from ward to neighborhood system. The Pingjiang map shows that Suzhou's old city is rectangular in plan, about 6 km north-south and 4.5 km east-west, with a formal and regular but free and flexible layout. Suzhou planners likewise applied the principles in the *Rituals of Zhou*, the Nine Squares System, and *Feng Shui*, making the city well integrated with the surrounding natural environment (*History of Chinese Architecture*, 1986/2009; Xu, 2000).

Pingjiang was laid out south by east 7°54' to take advantage of the southeast wind for cooling in the summer and avoiding direct winds in the winter. The eight city gates are in line with the eight trigrams (*bagua*) pointing to the eight principal directions as well as following the waterways (Xu, 2000; Yu, 2007).

---

10   At present, the amount of land requisition and demolition fees count for more than 50 percent of the total redevelopment cost in Beijing; among which the relocation fees account about 45 percent; whereas the land requisition and demolition fees in the new districts are much lower, accounting for only 14 percent of the total development cost (*Beijing City Planning Chart*, 2007).

11   相土尝水, 象天法地.

12   Plum rains (梅雨) are intermittent drizzles in the rainy season in the middle and lower reaches of the Yangzi River.

Suzhou[13] was planned with 'water' in mind since ancient Chinese looked upon water as blood of the earth and believed that the prosperity of a city depended on the circulation of water. The 14.2 sqkm old city has 35 km of artificial rivers, often 3-5 m deep and 10 m wide, directed from Lake Tai, because Suzhou is near the Yellow Sea in the east and the Yangzi River in the north; digging canals and building city walls were thought to be best strategies for flood prevention. Suzhou has a double-chessboard grid pattern, with waterways (*Yin*) and roads (*Yang*) complementing each other connected by 168 bridges. Principal canals run east-west and most streets are parallel to them. The urban blocks normally have a rectangular shape with 2-3-storey traditional houses along east-west lanes facing south-north. This arrangement forms a unique Suzhou style of 'little bridges, rivers, and houses.' This basic plan has been maintained in Suzhou's old city until today (Xu, 2000; Yu, 2007).

Suzhou was divided into 60 wards in the Tang dynasty (618-907) and 65 wards in the Song time (960-1279). With rapid economic development in mid-Northern Song (960-1127), a new neighborhood system emerged and gradually replaced the old ward system. The scroll painting *Suzhou Flourishing* (姑苏繁华图; figure 3.8) by Xu Yang in 1759 portrays the city's prosperity in the late 18th century and features the courtyard house system.

Suzhou was praised as 'Paradise on Earth' and 'City of Gardens' because of its favorable natural conditions for growing lush green woods, fish, and rice. With 1,500 years of gardening history, Suzhou had more than 170 gardens by the end of Qing (1644-1911), more than 60 of them are completely preserved, and 19 open to the public today. [14] Suzhou is such a place that "one can enjoy landscapes without going outside the city and live in busy streets with the sights of forests and tastes of spring water"[15] (my translation). Hence, in imperial China, retired officials, literati, and wealthy merchants settled in Suzhou, resulting in its prolific number of renowned celebrities and outstanding talents. Suzhou Pingtan is one of the top four national operas; story-telling, ballads singing, and Kun Opera are referred to as the 'Three Flowers of Suzhou Culture' (China Classical Tours, 2006; China Tourist Cities, 2002; ChinaTravel.com, n.d; Xu, 2000; Yu, 2007; Yuan and Gong, 2004; Zhou, 1998).

---

13   Suzhou covers an area of 8,488 sqkm with 1,650 sqkm of urban area. Overall, 10 percent is cultivated fields, 30 percent hills, and 42.5 percent waterway.

14   UNESCO World Heritage Committee has added nine classical Suzhou gardens to their list: *Zhuozhengyuan* (The Humble Administrator's Garden), *Liuyuan* (The Lingering Garden), *Wangshiyuan* (The Master-of-Nets Garden), *Huanxiushanzhuang* (Mountain Villa of Embraced Beauty), *Canglang Ting* (Surging Waves Pavilion), *Shizilin* (Lion Grove Garden), *Yipu* (Garden of Cultivation), *Ouyuan* (The Couple's Garden of Retreat), and *Tuisiyuan* (Retreat and Reflection Garden) (Yuan and Gong, 2004). Suzhou is nicknamed the 'Back Garden' of Shanghai as it is 83 km to the northwest of Shanghai, about 30 minutes by high-speed train.

15   不出城郭而获山水之怡, 身居闹市而有林泉之致.

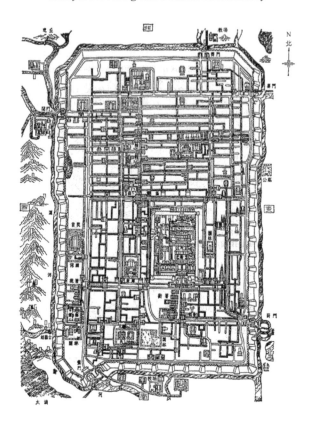

Figure 3.7   The Pingjiang map (dated 1229) from a stone stele showing a double-chessboard, road-river system. Source: *History of Chinese Architecture*, 1986/2009, p. 56

Suzhou's old city walls also suffered tragedy where more than half have been demolished since 1949, with the remaining 1442 m buried in houses and factories. In 2001-2005, the Suzhou municipal government rescued the city walls and made it a unique 'riverside green corridor' (Yu, 2007, pp. 301, 305-306).

Classical Courtyard House Design

When designing a courtyard house, the *Yellow Emperor's Canon of Internal Medicine* (黄帝内经) advises that five *xu* (虚 'deficiency') [16] will make its occupants poor and five *shi* (实 'fullness') will make them rich. The five *xu* include: a big house with fewer people; a small house with a big gate; incomplete enclosing

---

16   *Xu* (虚) and *shi* (实) is Chinese aesthetic concept that can be interpreted in numerous ways. *Xu* may denote "void, virtual, potential, unreal, intangible, formless, or deficient" and *shi* may mean "solid, actual, real, tangible, formed, or full."

Figure 3.8 **Partial scroll painting *Suzhou Flourishing* (姑苏繁华图) by Xu Yang in 1759 shows the courtyard house system. Courtesy of Liaoning Museum**

walls; the well and stove in wrong places; and a big plot with large courtyards but fewer rooms. The five *shi* include: a small house with many people; a large house with a small gate; complete enclosing walls; a small house with many livestock; and ditch water flowing to the southeast (Kou, 2005, pp. 59, 89-90; Ma, 1999, p. 38). Thus, in *Feng Shui* theory, it is important to match the size of a house and the size of a household.

The important halls and vital features of a house (e.g., gate, kitchen, stove, etc.) were positioned with the help of the verse formula *Bagua qizheng[17] da you nian*,[18] founded on the concept of 'cosmic resonance' (天人感应) that links heavenly stars (appendix, table A.1) with earthly houses to predict the auspiciousness or inauspiciousness of each space, thus generating eight types of residential orientations (figure 3.9).

The construction process took the following seven steps:

---

17   *qizheng* (七政) are the seven coordinate points casted by the sun along a reversed 'S' line, from which the *Yin Yang* symbol derived.

18   八卦七政大游年: 乾六天五祸绝延生; 坎五天生延绝祸六; 艮六五绝延祸生 天; 震六绝祸生延天五; 巽天五六祸生绝延; 离六五绝延生祸生天; 坤天延绝生祸五六; 兑生祸延绝六五天.

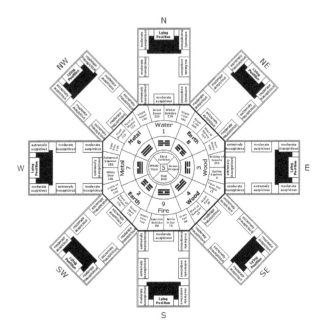

Figure 3.9   **An interpretation and translation of the *bagua* (eight trigrams)
method. Drawing by Donia Zhang based on Luò, 2006; Ma,
1999; Wang, 1999; Zou, 2002**

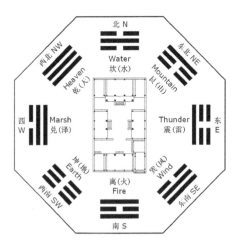

Figure 3.10   **Courtyard house and *bagua* implications. If taking the Northern
Hall as the lying position, the Northern Hall, Southern Hall,
Eastern Hall, and the southeast direction are all auspicious. If
taking the southeast corner gate as the lying position, the result
is the same. Drawing by Donia Zhang based on Ma, 1999, p. 41**

1. Divide the residence into nine squares and mark their orientations;
2. Create a central axis based on the owner's birthday and eight characters.[19] For ordinary houses, the central axis should not face direct south-north but off a little because commoners' fates are not strong enough to uphold south-north energy. Only the imperial palaces could face direct south-north orientation;
3. Locate the Central Hall in the 'lying position' (伏位) and turn clockwise to find the positions of the other stars (appendix, table A.1), or take the lying position for the gate, use the verse formula to organize the other rooms;
4. Determine the auspiciousness or inauspiciousness of each position based on their *Yin Yang* and *wuxing* of the stars (appendix, table A.1);
5. Build the room size, height, or number of floors according to the degree of auspiciousness/inauspiciousness of the position;
6. Drain the rainwater from east to west, or it would violate the 'peach blossom water,' which should be strictly avoided;
7. Use the 'Lu Ban Chi'[20] ruler to measure and cut materials for construction since the eight divisions in the central column help establish the favorable/unfavorable dimensions.[21] Take special care in measuring doors, windows, tables, and beds to ensure that their dimensions fall within the favorable range and hence bode well for the family (Knapp, 1999, p. 49; 2005d, pp. 114-117; Ma, 1999, p. 40; Wang, 1999, pp. 57, 90-93).

Harmony with Humans

> *'Straight' indicates the correctness (of the internal principle);*
> *'Square,' the righteousness (of the external act).*
> *The gentleman (thus represented),*
> *by his self-reverence maintains the inward (correctness),*
> *and in righteousness adjusts his external acts.*
>
> – *Yi Jing*, Appendix 4, Section 2 Clause 6

Harmony with humans is a central theme in Confucianism when we think of such advocated virtues as benevolence, righteousness, filial piety, fraternity, fidelity, dignity, loyalty, sincerity, obedience, forbearance, morality, responsibility,

---

19 生辰八字: birth year, month, day, and hour, each consists of one Heavenly Stem (天干) and Earthly Branch (地支).

20 'Lu Ban Chi' (鲁班尺) is also called 'Ba Zi Chi' (八字尺 'eight-character ruler') or 'Men Guang Chi' (门光尺 'door brightening ruler' because it was thought that cutting the sizes of the elements using this ruler could bring honor to one's ancestors 光宗耀祖).

21 For example, the first 财 ('wealth') and the last 吉 ('luck') are most favorable; the fourth 义 ('justice') and fifth 官 ('office') are moderately favorable, and the rest four: 病 ('illness'), 离 ('separation'), 劫 ('plunder'), and 害 ('harm') are unlucky.

modesty, honesty, trustworthiness, compassion, and the like (Lau, 1991, pp. 211, 222; Reid, 2000). These qualities form the basis of dealing with the five cardinal relationships that Confucius (551-479 BCE) identified: emperor-minister, father-son, husband-wife, elder-younger brothers, and friends (*Doctrine of the Mean*, Chapter 20 Section 8). Confucius promoted these values because he observed that a person with dignity would be met with respect, a person with compassion would be revered by the masses, a person with trustworthiness would have the confidence of the people, and a person with benevolence would be served willingly by others (*Analects*, Book 17 Chapter 6).

House Form and Human Behaviors

Confucianism suggests using geometric forms such as circles, squares, or straight lines in house design to regulate human behaviors inside, as Mengzi (Mencius, 372-289 BCE) wrote:

> When the sages had used the vigor of their eyes, they called in to their aid the compass, the square, the level, and the line, to make things square, round, level, and straight: the use of the instruments is inexhaustible...The compass and square produce perfect circles and squares. By the sages, the human relations are perfectly exhibited. (*The Works of Mencius*, Book 4 Part 1 Chapters 1 and 2)

Daoist influence on house and garden design is manifested in the interplay of open (*Yin*) and closed (*Yang*), and solid (*shi*) and void (*xu*) spaces (Knapp, 2005c). Daoism stresses the use of organic shapes to free people's imagination and creativity, as Zhuangzi (c.369-286 BCE) argued:

> If we must use curve and plumb line, compass and square to make something right, this means cutting away its inborn nature...For in the world there can be constant naturalness. Where there is constant naturalness, things are arced not by the use of the curve, straight not by the use of the plumb line, rounded not by compasses, squared not by T squares. (*The Complete Works of Chuang Tzu*, Chapter 8)

The Chinese found both philosophies complementary and valuable, and applied them side by side. While classical Chinese courtyard houses mirrored Confucian ethics to regulate human behaviors, classical Chinese gardens followed Daoist principle of harmony with nature (Keswick, 2003). However, as wooden-framed Chinese courtyard houses predated Daoism and Confucianism by almost 3,000 years (Qiao, 2002), it would be fair to say that these ideologies only refined the design concept at a later date.

Selection of Neighborhood/Neighbors

The Chinese have long held that good neighborhood location and neighborly relations are crucial for personal development. Confucius (551-479 BCE) advised people to associate with those who could advance you, and to put aside those who could not do so (*Analects*, Book 19 Chapter 3). It is virtuous manners that constitute the excellence of a housing estate, and a wise person should select a house where such qualities prevail (*Analects*, Book 4 Chapter 1).

There is a household legend in China about Mengzi's (372-289 BCE) mother who moved three times to choose the right neighborhood for him. Mengzi's father died when he was little. He and his mother lived near a cemetery. Mengzi often played the game of performing funeral ceremonies with other children, and the environment was unhealthy for him. They moved to a new home near a market. Mengzi then learned imitating merchants selling goods; he could not concentrate on his study. They moved once again to a new home near the Academy of Classical Learning. Mengzi began to learn obeying rules, following rituals, and enjoying study. Nurtured by the spirit of the Academy, Mengzi later became one of China's greatest thinkers, which could be attributed to his mother's wise choice of neighborhood for him (Luò, 2006; Yang, 2008).

Another account in the *Nanshi Lü Sengzhen Zhuan* (南史•吕僧珍传) records: "Ji Ya in the Song dynasty (960-1279) bought a house in the Nankang Prefecture, and the house is beside that of Sengzhen. Sengzhen asked about the price, Ji Ya replied: '11 million.' Sengzhen wondered why it was so expensive, Ji Ya answered: 'one million to buy the house and 10 million to buy the neighbor'" (Luò, 2006, p. 4; my translation). The story implies that good neighbors are hard to buy.

A popular phrase among Beijingers nowadays is that, "Distant relatives are not as dear as neighbors, and neighbors are not as dear as the people living opposite my door," revealing how important neighbors still are to contemporary Chinese (Bai, 2007).

Family Relations in Imperial China

A family is a basic unit of a society and the successful union of family members may advance a country. Harmony within family members was emphasized in imperial China as expressed in the idiom, "a harmonious family brings prosperity in all things" (家和万事兴).

Confucius prioritized filial piety towards one's parents and fraternity among siblings, because for him, filial piety and fraternal submission were the roots of all benevolent actions (*Analects*, Book 1 Chapter 2). Mengzi (372-289 BCE) then suggested that there were three things unfilial, and that to have no posterity was the greatest of them (Book 4 Part 1 Chapter 26). These sayings had profound influences on traditional Chinese family structure.

Despite the critical role of women in producing and nurturing a family's lineage, Confucius included an element on the inferiority of women by stating that girls and mean men are the most difficult to feed (*Analects,* Book 17 Chapter 25), which is principally incompatible with modern egalitarianism or liberalism. The modern Chinese request for equal rights for women has demanded a revision of this traditional thinking, just as in other parts of the world. Moreover, Confucian superiority of age to youth and the respect for hierarchical order can also cause problems in contemporary society. Nevertheless, Confucius must be understood in the socio-cultural context of his time if we attempt to comprehend his teachings in a meaningful way (Csongor, 1991, p. 452; Metzger, 1991, pp. 291, 300; Sprenger, 1991, p. 468).

In contrast, both philosophical and religious Daoism have their preferences for the female and *yin* force by giving a higher place for women, as Daoist writings frequently refer to 'Dao' as the 'Mother of all things' and show fondness for images such as water, valley, emptiness, and the like, over their opposites (Ching, 1993).

Harmony with Self

In imperial China, self-control rather than self-liberation was emphasized (Kubin, 1991, p. 69), and an active self-cultivation was mentioned as early as in *Yi Jing* (Yu, 1991, p. 53) and later promoted by Confucians and Daoists.

Self-Cultivation

Confucius (551-479 BCE) argued that by nature, humans were nearly alike, but by practice, they got to be wide apart (*Analects*, Book 17 Chapter 2). When a person cultivates to the utmost the principles of his/her nature, and exercises them on the principle of reciprocity, s/he is not far from the path (*Doctrine of the Mean*, Chapter 13 Section 3). Confucius also contended that to restrain the self and return to propriety was perfect virtue. The step to it is not to speak or act contrary to propriety. The person of perfect virtue is cautious in his/her speech as for one word a person is often deemed to be wise, and for one word s/he is often thought to be foolish (*Analects*, Book 12 Chapters 1 and 3, Book 19 Chapter 25).

Confucius considered serving the state as a moral obligation of scholars and suggested:

> The ancients who wished to illustrate illustrious virtue throughout the kingdom, first ordered well their own states. Wishing to order well their states, they first regulated their families. Wishing to regulate their families, they first cultivated their persons. Wishing to cultivate their persons, they first rectified their hearts. Wishing to rectify their hearts, they first sought to be sincere in their thoughts. Wishing to be sincere in their thoughts, they first extended to the utmost their knowledge. Such extension of knowledge lay in the investigation of things. (*Great Learning*)

Thus, a perfect Confucian personality *junzi* ('gentleman') cultivates his desire for learning.

Being Lived by the Dao

Whereas for Laozi (c.571-471 BCE), self-cultivation indicated an individual's harmonious relationship with the Dao – the laws of nature, and his naturalistic approach is expressed throughout *Dao De Jing* (The Book of the Way), for example, "The Master keeps her mind always at one with the Dao" (Verse 21) because "Only in being lived by the Dao can you be truly yourself" (Verse 22) and "If you open yourself to the Dao, you are at one with the Dao" (Verse 23), "If powerful men and women could remain centered in the Dao, all things would be in harmony" (Verse 32), and that they may "spend weekends working in their gardens, [and] delight in the doings of the neighborhood" (Verse 80).

Daoism advocates respecting the natural environment and preserving the natural order since nature is both inside and outside us and each person embodies the Dao. Harmony with self also implies harmony with our own body. Natural healing methods such as acupuncture and herbal remedies, seasonal diets, and respiration exercises (*taiji* and *qigong*), all aim to stimulate and balance the flow of *qi* so that human body systems could work together in harmony. Once a balanced state of health is achieved, the body radiates and extends harmony to the outside world. Thus, to regulate the world, we must first cultivate ourselves, to tend our inner and outer garden, and when someone helps a garden grow, this continuous, holistic, and dynamic process of development deepens the gardener's understanding of nature and transforms his/her life profoundly, as it is commonly acknowledged that a healthy environment nurtures a healthy person (Kohn, 2001, pp. 382-386; Schipper, 2001, pp. 91-92).

Central Harmony

One of the most important Daoist texts *Tai Ping Jing* (The Book of Great Peace) of the Eastern Han dynasty (25-220 CE) suggests Central Harmony as a way to establish harmonious relationships among Heaven, Earth, and humans, and the formation of the state of Central Harmony will come the era of Great Peace (*taiping*) or Great Harmony (*taihe*). Daoist Central Harmony echoes Confucian Middle Way in the *Doctrine of the Mean*, and is meant to expand the bipolar theory of *Yin Yang* and formulate a theory of triad: great *yang* (*taiyang*), great *yin* (*taiyin*), and Central Harmony (*zhong he*). Humans belong to the realm of Central Harmony and are charged with the responsibility of conserving, circulating, and cultivating the *qi* to maintain a harmonious union of Heaven and Earth and to bring about cosmic harmony and social peace (Lai, 2001, p. 96). The centrally positioned courtyard in a house compound connects Heaven, Earth, and humans and facilitates the flow of *qi*.

Neo-Confucian philosopher Zhu Xi (1130-1200) considered that the circulation of *qi*, *yin yang*, and that of the four seasons produced and sustained the order of life, and the natural order synchronized with space (east, south, west, and north) and time (hours, days, months, and years). In Neo-Confucianism (merged Confucianism and Daoism and formulated in the 11th-12th centuries), time is characterized as a circle of the Five Phases of *qi*. Patterns such as trees growing and flowers blossoming are the four aspects of life: spring is its generation, summer is its growth, autumn is its completion, and winter is its storage. These four phases of life correlate to the continuous functions of time and the collapse of these cyclical functions would result in the failure to sustain lives. For example, when the cycle is disrupted, lives are not generated in spring (Kuwako, 1998, pp. 162-163; Weller and Bol, 1998, p. 321). As such, a way of honoring the natural order is by celebrating seasonal change, which one may comprehend less through words than by immersing him/herself in the rhythms of nature (Knowles, 1998, 1999; Neville, 1998, pp. 266-267).

Winter Solstice

Autumn Equinox   *Spring Equinox*

Summer Solstice

Figure 3.11   Proposed *Yin Yang Yuan* (阴阳元[22]) symbol depicting gradual, cyclic movements of sunlight and the four seasons. It expands traditional *Yin Yang* symbol (figure 3.1) by offering clearer demarcations of the start of each season, and highlights triadic and quadric than dualistic divisions. Design and drawing by Donia Zhang 2009

---

22   元 (*Yuan*) is a Chinese philosophical concept, referring to the basis for the world's existence, the Dao – the way things follow their natural course, as celestial bodies all follow their own tracks. For example, the phrase 一元复始 refers to the beginning of the New Year. Nevertheless, Chinese dictionaries list several meanings of 元: its first meaning is beginning, primary, head, or origin; its second meaning is basic or fundamental; and its third meaning is a unit. The Kangxi Dictionary《康熙字典》says: "天地之大德，所以生生者也。元字从二从人，仁字从人从二。在天为元，在人为仁，在人身则为体之长。" (Retrieved November 6, 2011 from http://baike.baidu.com/view/6673.htm)

My conceived *Yin Yang Yuan* (阴阳元) symbol (figure 3.11) is an expansion of traditional *Yin Yang* symbol (figure 3.1) aiming to overcome the philosophical dilemma of binary thinking to avoid extremities. It depicts gradual, cyclic movements of sunlight and is supported by the concepts of Central Harmony in Daoism and the Middle Way in Confucianism. It relates to the study in that a courtyard house is also termed 'quadrangle' because of its four-sided enclosure, and that the courtyard at the center allows direct observations of light changes throughout the year that lead to seasonal festivities: Spring Festival, Midsummer/ Summer Solstice Festival (Dragon Boat Festival), Mid-Autumn Festival, and Winter Solstice Festival (Chapter 5).

Table 3.1   Connotations of *Yin Yang Yuan* symbol

Yin Yang Yuan Triad	<ul><li>Past-Present-Future</li><li>Sun-Moon-Stars</li><li>Heaven-Earth-Humans</li><li>South Pole – North Pole – Equator</li><li>Morning-Afternoon-Evening</li><li>Day-Night-Twilight</li><li>Up-Down-Middle</li><li>Hot-Cold-Temperate</li><li>Dry-Wet-Moist</li><li>Interior-Exterior-Transitional space</li><li>Positive-Negative-Neutral</li><li>Active-Inactive-Passive</li><li>Maximum-Minimum-Optimum</li><li>Yes-No-Unsure</li><li>Hard-Soft-Moderate</li><li>Mountains-Rivers-Plains</li><li>Father-Mother-Child</li></ul>
Yin Yang Yuan Quadrant	<ul><li>Spring-Summer-Autumn-Winter</li><li>Sun-Moon-Stars-Planets</li><li>Heaven-Earth-Humans-Self</li><li>South-North-East-West</li><li>Morning-Afternoon-Evening-Night</li><li>Dawn-Midday-Dusk-Midnight</li><li>Up-Down-Left-Right</li><li>Father-Mother-Son-Daughter</li></ul>

Source: My summary

Thus, both Confucianism and Daoism emphasized self-cultivation to establish a harmonious relationship with the natural and social environments. These ideologies formed the basis of Chinese culture for two millennia and built a solid foundation for the stability of imperial China until the first Anglo-Chinese War (Opium War, 1839-1842) destroyed China's social harmony. This part of history is depicted in the film 火烧圆明园 (*The Burning of Yuan Ming Yuan*, 1983) showing

British and French troops invading Beijing and burning the imperial Winter Palace in the western suburb and killing many Chinese people at the time.

Summary and Conclusion

This chapter offered an in-depth literature review on relevant parts of Chinese philosophy, particularly *Yin Yang* and how it fits into *Yi Jing* and *Feng Shui* theories that have helped to plan and design imperial Chinese capital cities and classical courtyard houses.

Yi Jing and *Feng Shui* were generated from the notion of 'unity of heaven and humans' and 'cosmic resonance,' a cosmology that suggests that to succeed, all human activities on earth should follow the Dao – the natural laws. *Feng Shui* is practical; it is supposedly based on objective, factual, and empirical observations of the aesthetic order of nature (Anderson, 2001, pp. 172-173; Field, 2001, p. 187; Kou, 2005; Luò, 2006) and its ultimate purpose is to create favorable living conditions conducive to the occupants' health and wellbeing.

The chapter has revealed that the fundamentals of Chinese philosophy are to establish harmony on four different levels: within the individual (harmony with self), between the individual and the society (harmony with humans), between the individual and the natural environment (harmony with earth), and between the individual and the universe (harmony with heaven). Humans are also responsible for harmonizing Heaven and Earth (central harmony). The ideal degree of attainment is the integration of oneself with the natural and social environments in a sense of cosmic oneness, which will require continuous efforts to grow and refine oneself (Kohn, 2001, p. 389; Kubin, 1991, pp. 76, 82; Tu, 1998b, p. 116).

Confucian and Daoist thought has profoundly influenced traditional Chinese family structure and classical courtyard house layout because these values were not just philosophical knowledge that existed only in books, but rudimentary rules that governed everyone's daily life (Weggel, 1991, p. 407). Since human nature does not seem to have changed very much over the last two millennia, some Confucian and Daoist values may well serve the contemporary world when it is characterized by so many conflicts.

The proposed *Yin Yang Yuan* symbol is an expansion of the traditional *Yin Yang* symbol that suggests the idea of central harmony or a middle way when dealing with matters; it also recommends a gradual change when encountering questions such as cultural or urban transformation. Moreover, it implies the preservation of the natural order and honoring nature's cycle.

In Chapter 4 that follows, the form and environmental quality, and space and construction quality of classical Chinese courtyard houses will be discussed to provide the context for the study.

PART II
Classical Courtyard Houses of China

Chapter 4
Form and Environmental Quality, Space and Construction Quality of Classical Courtyard Houses

The world is formed from the void,
like utensils from a block of wood.

– Laozi, c.571-471 BCE, *Dao De Jing*, Verse 28

This is one of the three background chapters that place courtyard housing in its historic and symbolic context to explore the architectural and spatial design elements of classical Chinese courtyard houses in Beijing and Suzhou, and to investigate the cultural continuity of the renewed/new courtyard housing examined in Chapters 7-8. The chapter consists of 11 sections: exterior form, exterior walls, gate and access, windows, courtyards and gardens, roofs, interior space, floor levels, furniture styles and materials, facility provision, and building materials and construction quality.

Exterior Form

Beijing's basic urban unit was the classical courtyard house (*siheyuan*) in the residential quarters of the old city. Urban form determined the shape and size of *siheyuan*. Beijing during the Yuan dynasty (1279-1368) had 50 neighborhoods (*fang*) and three types of roads. The width of thoroughfares measured about 37 m, streets around 19 m, and *hutong*[1] ('lanes') 6-9 m. Streets ran mostly north-south and *hutong* east-west. The distance between two *hutong* was 77 m, which allowed a house to have either two courtyards with a garden, or three courtyards. A *hutong* of 440 m long could accommodate 7 *siheyuan*, but many aristocrats built their mansions across two *hutong* (Bai, 2007, p. 9; Wang, 1999, p. 23; Wu, 1999, p. 77).

The wealthy land buyers were each allocated 8 *mu* (5333 sqm) in Beijing's inner city; the plots determined the typically square-shaped *siheyuan*. During the Ming

---

1   A *hutong* is a narrow lane with houses along each side. The term is believed to be Mongolian 'hottog' in origin, since it first appears in Beijing's Yuan dynasty records; there are several theories about the original meaning of the word, the most popular being that it means 'water well' (Wang, 1999, p. 23; Wu, 1999, p. 69).

(1368-1644) and Qing (1644-1911) dynasties, Beijing *siheyuan* exhibited clear variations, which could be attributed to the influx of the Măn ethic minority who took power and led to an increase in population and change in house structures. The land designated to each household was reduced from 8 *mu* to 1-5 *mu* (667-3333 sqm). The front of the courtyard house compound measured between 15-25 m and the depth 25-50 m. Some of Qing *siheyuan* are still in use today (Alexander *et al.*, 2003; Bai, 2007, pp. 4, 9-10; Knapp, 2000, 2005a, 2005c; Ma, 1999).

Originally designed to house one extended family, a fully developed *siheyuan* had four basic characteristics: introverted form, central axis, bilateral symmetry, and hierarchical structure (Casault, 1987, 1988; Knapp, 2000, 2005a; Liang, 1998). A *jin* (进 'rise') comprised a hall and a courtyard along the north-south central axis. There were five basic types of *siheyuan* in Beijing:

1. One-courtyard house (sometimes 'three-sided enclosure' called *sanheyuan* without north-facing hall; figures 4.1 and 4.2);
2. Two-courtyard house (small-size; figure 4.3);
3. Three-courtyard house (typical, standard, or medium-size; figure 4.4);
4. Four-courtyard house (large-size; figure 4.5);
5. Five-courtyard house with a garden (large-size; figure 4.6).

Other less common types included: one primary and one secondary courtyard (跨院) compounds built side by side, and two- or multi-group courtyard compounds built side by side. In feudal China, some large families had two brothers living in the same household. Thus they constructed two compounds side by side of the same or similar size and layout, independent yet connected (Ma, 1999). Beijing during the Ming dynasty (1368-1644) prohibited commoners from building a house of more than three courtyards unless they were an official. This regulation was somewhat loosened during the Qing dynasty (1644-1911) (Chan and Xiong, 2007). Some medium-sized courtyard houses also had a carriage (or sedan chair) room in the front (figure 4.7) adjacent to the *hutong* (Bai, 2007; Ma, 1993, 1999).

The old city of Suzhou had north-south and east-west running streets that were 200-400 m in length, and the distances between the blocks were 60-80 m, within which traditional houses and gardens were built. Two lanes were connected by houses that had a number of small courtyards/lightwells, some with small gardens; the compounds were enclosed by high walls (resembling Beijing *siheyuan*). For river-side houses, the river and street were parallel with a row of houses in between, or the river in the middle of two rows of houses. The aesthetic principle of 'small is beautiful' was well implemented in Suzhou; everything was built on a human scale. In a traditional Suzhou house, a *jin* (进 'rise') was comprised of a hall and a courtyard or lightwell along the north-south main axis. A *luo* (落 'set') was consisted of halls and courtyards/lightwells along a side axis. The halls along the main axis were termed *zhengluo* (正落 'main set'), and those on the side axes were called *bianluo* (边落 'side set'), or *ciluo* (次落 'secondary set'). The rises and sets

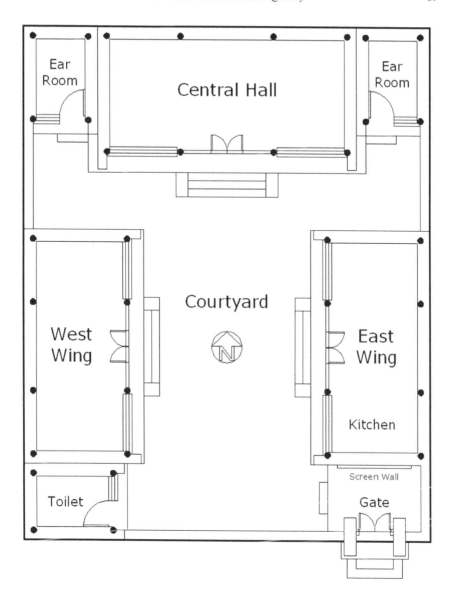

Figure 4.1   Plan of three-sided (*sanheyuan*) one-courtyard house of Beijing.
Drawing by Donia Zhang after Ma, 1999, p. 15

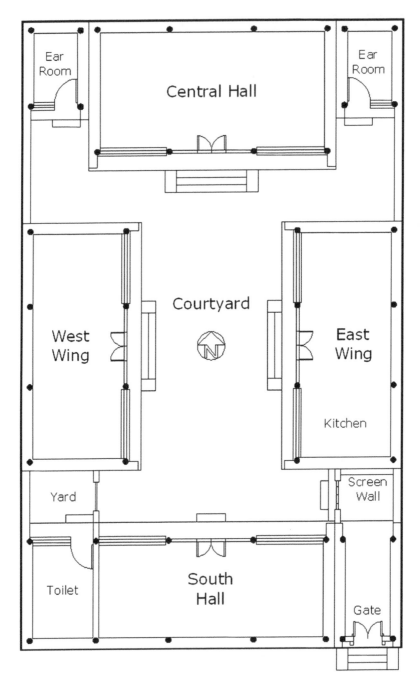

**Figure 4.2   Plan of four-sided (*siheyuan*) one-courtyard house of Beijing.
Drawing by Donia Zhang after Ma, 1999, p. 15**

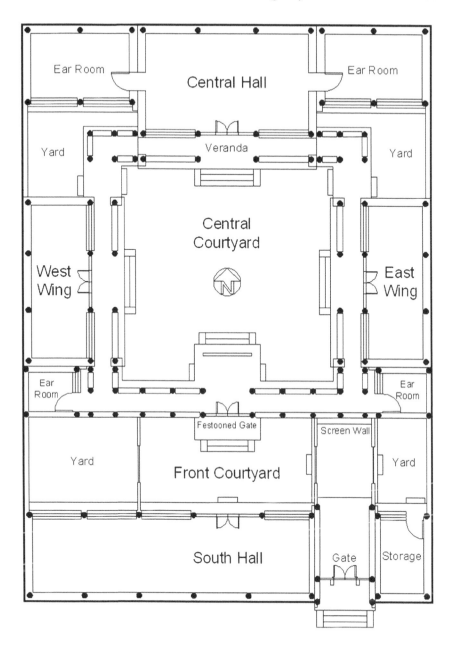

Figure 4.3 Plan of two-courtyard house of Beijing. Drawing by Donia Zhang after Ma, 1999, p. 17

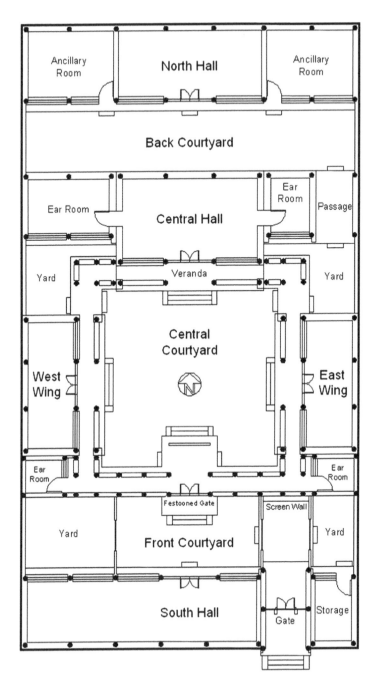

Figure 4.4   Plan of three-courtyard house of Beijing. Drawing by Donia Zhang after Ma, 1999, p. 17

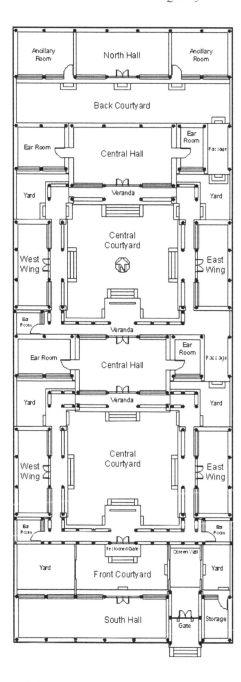

Figure 4.5   Plan of four-courtyard house of Beijing. Drawing by Donia Zhang based on Blaser, 1995, p. 19; Ma, 1999, p. 19; Chan and Xiong, 2007, p. 45

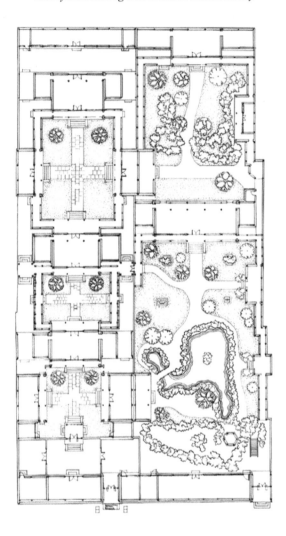

Figure 4.6   Plan of five-courtyard house with gardens, Beijing. Source: Ma, 1999, p. 25

Figure 4.7   Model of a typical three-courtyard house of Beijing with a carriage room (or garage) in the front. Source: Ma, 1999, p. 9

Figure 4.8   Model of a three-courtyard house of Beijing with a garden in the east. Source: Ma, 1999, p. 49

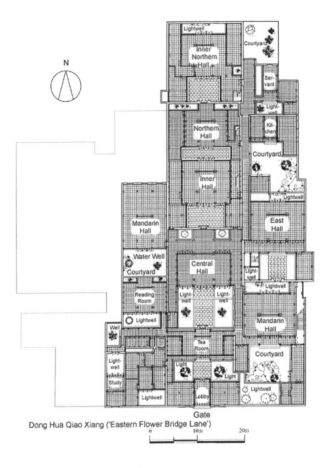

Dong Hua Qiao Xiang ('Eastern Flower Bridge Lane')

Figure 4.9   Plan of Wang Mansion, Eastern Flower Bridge Lane (东花桥巷), **Suzhou. Source: my translation from Suzhou Housing Management Bureau, 2004, p. 62**

created a spatial rhythm and series of contrasts between *Yin Yang* and *xu* (void) *shi* (solid) (Yu, 2007). There were generally four types of traditional Suzhou houses:

1. Small house with 1-2 courtyards/lightwells along the north-south main axis;
2. Medium house with 2-3 courtyards/lightwells along the main axis;
3. Large house with 5-7 (occasionally 10-20) courtyards/lightwells surrounded by halls along the main axis, and several 10s of courtyards/lightwells along the side axes;
4. Extravagant garden house estate with pavilions, ponds, rockeries, trees, and flowers (origins of Suzhou gardens) (Yu, 2007).

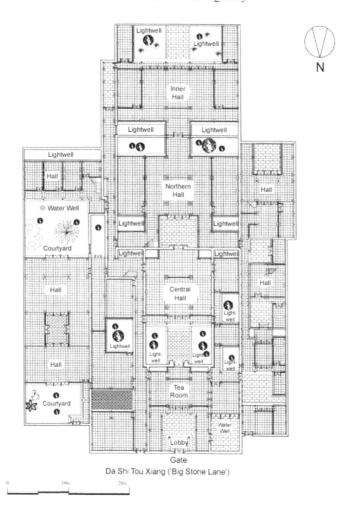

N

Figure 4.10   Plan of Wu Mansion, it was the residence of Qing dynasty writer Shen Fu (沈复, 1763-1810?), author of *Six Chapters of a Floating Life* (浮生六记); later a Wu-surnamed family bought the house, Big Stone Lane (大石头巷), Suzhou. Source: my translation from Suzhou Housing Management Bureau, 2004, p. 92

The tight plans of small Suzhou houses were commonly rectangular (longer in the east-west direction to reduce summer sunlight), T-shaped, three- or four-sided enclosures. Medium Suzhou houses normally had a depth of 70-80 m fitting in two (south and north) households. The house in the south usually had 3-4 courtyards/ lightwells, and the house in the north had 2-3 courtyards/lightwells. The medium Suzhou houses often had one set without a sidewalk. The layout was similar to the large houses except that there were fewer ancillary rooms and no garden; the lobby

Figure 4.11   Redeveloped Suzhou courtyard houses at Jiangfengyuan housing estate. Photo by Donia Zhang 2007

Figure 4.12   Model of a small riverside courtyard house in Suzhou Folk Custom Exhibition Center. Photo by Donia Zhang 2007

**Figure 4.13   Warm bridge connecting two riverside houses, Pingjiang
Historic District, Suzhou. Photo by Donia Zhang 2007**

and the inner/upper hall were also smaller, and the street façade was narrower and
simpler. Nonetheless, the enclosing walls and roofs had level changes, resulting in
an interesting profile (Yu, 2007).

Riverside Suzhou houses were generally small (figure 4.12), with many styles
of water footrests and water piers outside the house. To make more space, multi-
storey houses often had setbacks on the ground level and cantilevers above on the
riverside, forming a lively alteration in building height and depth. Some houses
were built across a river with a 'warm bridge' (figure 4.13) linking two rooms or a
rear gate so that the occupants could reach either side of the lane (Yu, 2007).

Exterior Walls

Beijing *siheyuan* had grey exterior walls for commoners and red walls for officials,
red (*Yang*) beams and columns, and green (*Yin*) window and door sashes (Ma, 1999).
Traditional Suzhou houses had white exterior walls, brown beams and columns,
and brown door and window sashes. The exterior walls in Suzhou were more solid
than those of Beijing to insulate from the intensive summer heat and winter cold,
since the houses traditionally did not provide a heating system for winter. The
high enclosing walls of classical Chinese courtyard houses often had assorted/

lattice windows pierced in a variety of organic or geometric patterns (e.g., circular, square, rectangular, fan, flower) for sunlight and air to pass through and so that one could have a glimpse of the sceneries inside the walls. The assorted windows were artistically designed and carved, making the house compound feel as if cut off from the street yet not (Yu, 2007).

Gate and Access

A Beijing *siheyuan* was typically accessed through a main gate at the southeast corner, and sometimes a rear gate at the northwest corner. In imperial China, the design and decoration of house gate was an important sign of the rank or wealth of the owner. Although there was a difference in setback distances and grandeur, the gate should always be double-leaf (1.8-2 m wide and 1.9 m high) (Bai, 2007; Ma, 1999). Upon entering a *siheyuan*, one would observe a screen/spirit wall facing the gate to protect the privacy of the household and eschew 'evil.' For wealthier families, a second 'festooned gate' (2.5-3 m wide and 2.8-3.6 m deep) was often built between the outer and inner courtyards. Some owners occasionally used the festooned gate as a temporary stage for artistic performances (e.g., Peking opera, a form of traditional Chinese theater which combines music, vocal performance, mime, dance, and acrobatics). The festooned gate thus delineated the semi-private and private zones of the house; it also marked the demarcation for female family members when greeting or sending their visitors. A common restriction for maidens was "not to step out of the second festooned gate"[2] (Bai, 2007; Ho, 2003; Ma, 1999). Covered verandas between 1.3-1.6 m wide and 2.2-2.4 m high were often built around courtyards as circulation and transitional spaces between indoors and outdoors (Ma, 1999).

Traditional Suzhou houses normally had two gates. The front/main gate was usually positioned at the center of the south wall, primarily used for daily activities, and the rear gate for services. Regardless of the shape of the site, a house typically had a clear longitudinal north-south main axis from the front gate through to the rear hall and functioning as the ceremonial access reserved for important occasions. On ordinary days, the doors to the halls along the main axis were locked. Daily access to the halls was through the sidewalks called *beinong* (备弄, 1.2-2 m wide), which were verandas for circulation and connecting different parts of the complex, and buffer zones for fire prevention, thus they were also called 'fire lanes' (火巷 *huoxiang*). *Beinong* could also help separate male and female rooms, and master and servant rooms. Hence they were sometimes called 'alleys of avoidance' (避弄 *binong*). In the late Qing dynasty (1644-1911), the sidewalks evolved from single-storey to 2-storey, horizontally and vertically connecting the whole compound. There were rules regulating which sidewalks the servants could (or could not) use. Visitors were greeted using different sidewalks according to their rank or relationship with the owner (Yu, 2007).

---

2   大门不出，二门不迈。

Figure 4.14 **The most common type Ruyi (如意 'as you wish') Gate of Beijing**
siheyuan. **Source: Ma, 1999, p. 195**

Figure 4.15 Suiqiang (随墙门 'along the wall') Gate of Beijing *siheyuan*.
Source: Ma, 1999, p. 195

Figure 4.16   Guangliang (广亮大门 'wide and bright') Gate of Beijing *siheyuan* with two screen/spirit walls outside to accentuate the higher social status of the owner. Source: Ma, 1999, p. 195

Figure 4.17   The 'festooned gate' of Prince Gong's Manor viewed from the front courtyard, Beijing. Photo by Junmin Zhang 2010

Figure 4.18 A typical gate with brick-carvings at Pan's Mansion, Weidaoguanqian Lane (卫道观前), Suzhou. Source: Suzhou Housing Management Bureau, 2004, p. 33

Figure 4.19   The most excellent gate brick-carving example at Master-of-Nets Garden in Suzhou with a height of about 6 m and the width of the carvings 3.2 m. The four engraved characters connote good fortune, emolument, longevity, and joy (*Explore*, October 2007, p. 84). Image source: Shao, 2005, p. 88

Figure 4.20   Gate and lattice windows viewed from the main courtyard, Suzhou. Source: Suzhou Housing Management Bureau, 2004, p. 38

Windows

In northern China, the Central Halls (正房, or Middle Halls, Ancestral Halls, Main Halls) were always 'sitting north and facing south' along the central axis for receiving sunlight. The rest of the rooms were orientated towards east and west. Efforts were often made to block cold winter winds by eliminating windows and doors on the back and side walls (Ma, 1999), but there were exceptions. In southern China, traditional houses were generally orientated towards north-south as there were rarely any east- or west-facing windows in Suzhou, and more attention was paid to ventilation and controlling the amount of sunlight (Knapp, 2005a; Liu, 2002a; Qiao, 2002).

Although classical Chinese courtyard houses created pleasurable living environments, their fundamental disadvantage was that all the major facades faced the courtyard, displaying only walls with hardly any openings to the public realm of lanes (*hutong* or *xiang*), and thus a less active street frontage.

Figure 4.21   Windows of the Central Hall facing the courtyard, a typical classical courtyard house, Beijing. Photo by Donia Zhang 1991

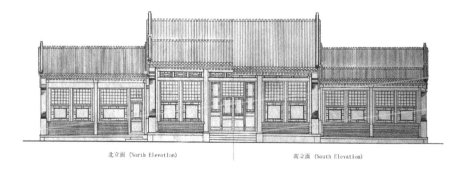

北立面 (North Elevation)   南立面 (South Elevation)

Figure 4.22   Elevation of the front/south hall and its two ear rooms, a Princess' Mansion, Beijing. Drawing by Donia Zhang 1991

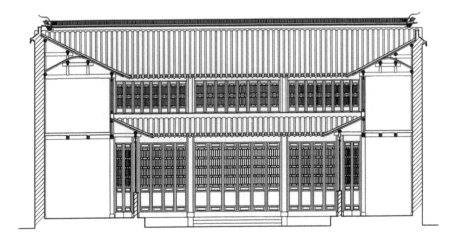

Figure 4.23   Elevation of the Inner/Northern Hall, Wang Mansion, Eastern Flower Bridge Lane (东花桥巷), Suzhou. Source: Suzhou Housing Management Bureau, 2004, p. 64

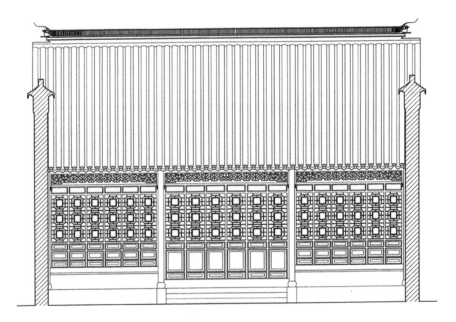

Figure 4.24   Elevation of the Central Hall, Wu Mansion, Big Stone Lane (大石头巷), Suzhou. Source: Suzhou Housing Management Bureau, 2004, p. 97

Courtyards and Gardens

Breezes in spring,
Flowers in summer,
Moon in autumn,
Snow in winter.[3]

      – Chinese poem on life in a courtyard garden; cited in Keswick, 2003, p. 199

Generally, in the hot and humid regions of southern China, the proportion of open to closed spaces decreases significantly compared to the cold and dry north and northeast where open spaces increase extensively (Knapp, 2000, 2005a, 2005c).

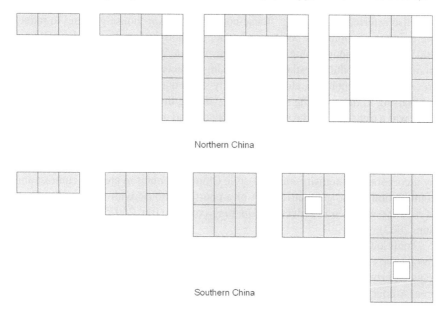

Northern China

Southern China

Figure 4.25 **Comparison of courtyard evolution and proportion in northern and southern China showing northern courtyards are generally larger than those of southern China. Drawing by Donia Zhang after Huang *et al.*, 1992, p. 5; adapted from Ronald Knapp (2005a, p. 25) in *Chinese Houses* by Tuttle Publishing**

In northern China, the courtyards were enlarged in the south-north direction for receiving more sunlight in winter. This layout mainly originated in the middle reaches of the Yellow River (Knapp, 2005a). Beijing's traditional courtyards occupied about 40 percent of the total ground area with the courtyard dimension

---

3   春风, 夏花, 秋月, 冬雪.

ranging from 7-20 m (Alexander *et al.*, 2003; Knapp, 2000, 2005a, 2005c; Ma, 1999), and the proportion of building height: width: length was 1:3:4 or 1:3:5 (Zhang, 1994, 2006, 2009/2010/2011).

The courtyard as an 'outdoor parlor' was often paved with bricks for domestic activities. Although the courtyard was a pleasant outdoor space, the occupants had to walk across it to reach each room even in severe weather conditions, an inconvenience especially for the elderly. In summer or during a wedding ceremony, a canopy was often erected in the courtyard as an extension of the Central Hall to accommodate more guests (Wang, 1999, p. 118).

In China, traditionally, gardens were an integral part of a scholar's house to the extent that the Chinese concept of a home was explicitly expressed in the term *yuanzhai*, meaning 'garden-house' (Knapp, 2005a). A house/home, or *jia*, was also referred to as *jiating* 'house-court,' or *jiayuan* 'house-garden,' to reflect the intimate relationships between a house, its courtyard, and its gardens (Wang, 2005, p. 75). Although some garden-houses were large estates, most mimic nature in miniature to exhibit aesthetically pleasing aspects of complementary elements (rocks, water, plantations, etc.) in relatively small outdoor spaces (Knapp, 2005a).

Starting in the pre-Qing and especially during the Qing dynasty (1644-1911), there was a trend among the wealthy merchants and officials in Beijing to integrate their courtyard houses with gardens. Efforts were made to create the most interesting views in tight urban spaces so that the gardens could be enjoyed daily and directly. Some Beijing mansions were further developed into garden houses by incorporating exquisite southern-style Chinese gardens to create a contrast and combination of the geometric form of a courtyard house with the organic shape of a Chinese garden to reflect the *Yin Yang* concept (Bai, 2007; Ma, 1993, 1999; Sun, 2002).

Traditional Suzhou houses often had both courtyards and lightwells, and in certain parts of southern China, the tiny open spaces of lightwells were still referred to as 'courtyards.' The total openings seldom exceeded 20 percent of the ground area. A courtyard/lightwell was placed in the front and back of each hall, and a larger courtyard was often located in front of the Central Hall to accommodate more guests as situation arose (Yu, 2007). The lightwells were generally a few square meters, and the smallest ones only 1-2 sqm, nicknamed 'crab eye lightwell' (蟹眼天井), where a cistern was usually placed to catch rainwater. The small lightwells helped reduce direct solar gain in summer, meanwhile producing strong wind suction like chimneys, forming a microcirculation within the house compound. Hence, even in the hottest months, such a house would make one feel reasonably cool. Tall plantations (bamboos) were sometimes grown in the lightwells. The disadvantage was that the narrow lightwells received poor daylight and often gave a sense of denseness (Knapp, 2000, 2005a; Liu, 2002a; Qiao, 2002).

In Suzhou during the imperial era, young people mostly received education at home from a private tutor. Some houses designated a courtyard solely for study. The study courtyard was often foursquare-shaped, about 1-2 times larger than an ordinary lightwell, rarely over 100 sqm, where one could take a leisurely walk,

Figure 4.26 The south/central courtyard, a typical classical courtyard house, Beijing. Photo by Donia Zhang 1991

Figure 4.27 The north/back courtyard, a typical classical courtyard house, Beijing. Photo by Donia Zhang 1991

Figure 4.28   A courtyard in a classical courtyard house in Shichahai Lake District, Beijing. Photo by Donia Zhang 2007

Figure 4.29   Suzhou lightwell for light, air, and rainwater. Photo from Lynn Xu 2009

Figure 4.30   Planting inside Suzhou lightwell. Photo from Lynn Xu 2009

read, or paint, accompanied by the sceneries of the four seasons. Large Suzhou houses usually had gardens on one side or at the back, which were extensions of the interior space. Large Suzhou gardens directed water from the rivers in the city without using a sluice gate, so that it would drain water in case of heavy rains in summer to protect people from humidity, prevent flooding, and reduce the effect of seasonal drought. Most garden ponds relied on groundwater, connecting the pond with the well water. Even in time of drought, the pond would not desiccate. Also, the well water was warm in winter, good for breeding fish (Suzhou Housing Management Bureau, 2004; Yu, 2007).

Gardens were the most welcoming spaces in Suzhou houses because they were lively, quiet, and tasteful. The owners of Suzhou houses could enjoy communication with Heaven and Earth in their small gardens. Suzhou gardens were ideal environments for scholars, painters, and officials to retire from public life and live in seclusion. Rich landscaping and superb spatial art impelled scholars and painters to maintain their good mood and nourish their imagination, which gave rise to their passion for creativity and resulted in many masterpieces. Thus, the secluded scholars naturally favored these gardens. If the 'culture of seclusion' was a unique phenomenon in old China, Suzhou gardens were important embodiments and outstanding representations of this culture. These gardens had intelligently combined nature and culture, creating 'poetic sentiments and artistic conceptions' (诗情画意) (Wang, 2005, p. 89; Yu, 2007, pp. 216-223).

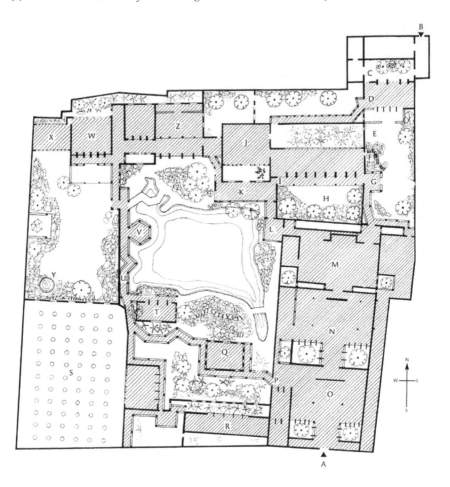

Figure 4.31   Plan of the Master-of-Nets Garden that used to be Li Mansion.
              The Garden is part of one of the best-preserved Suzhou houses.
              Chinese feudal societies had strict rules that women should
              not go out at will; thus the courtyards and gardens may have
              represented their entire world. From *The Chinese Garden* by
              Maggie Keswick. Published by Frances Lincoln Ltd. (1986)

Figure 4.32   Master-of-Nets Garden, Suzhou. Photo by Donia Zhang 1995

Figure 4.33   Master-of-Nets Garden, Suzhou. Photo by Donia Zhang 2007

Roofs

The most striking feature of a classical Chinese courtyard house was the elaborate roof. Since the Chinese worshipped Heaven, they employed a large, sweeping overhang as a link between Heaven and Earth, while at the same time expressing their aspirations to enter an eternal life (Liu, 1989). Large eaves also helped protect the walls and wooden columns from rainwater while allowing unobstructed daylight. There were many regional variations of roof types and degree of slope, depending on the temperature, rainfall, winds, and availability of material (mud compositions, thatch, clay tiles, wood, and stone shingles, etc.) (Knapp, 2005a; Liang, 1998).

Traditionally, a tiled pitched roof was the norm for buildings in eastern China, and the degree of slope was at least 4:12, often between 6:12 and 8:12 (or the inclination angle at least 30°, normally between 45°- 60°) (*Characteristics of Timber-Structured Chinese Ancient Buildings*, 2008). The depth of eaves was typically 60 cm (or 1:6 in northern China) to prevent rainwater from slanting in.

Figure 4.34   Roof detail. Pencil drawing by Donia Zhang 1991

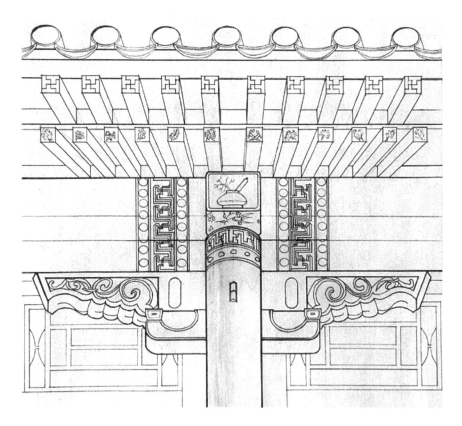

Figure 4.35   Roof and veranda detail. Pencil drawing by Donia Zhang 1991

Although the double-pitched roof was the most common, single-pitched, flat, and even convex-curved roof profiles were sometimes built in northern and north-eastern China. Examples showed that classical Chinese courtyard houses combined pitched and flat roofs in the same complex (Knapp, 2005a; Ma, 1999), but flat roofs were used only for service rooms. In southern provinces, gable walls often rose above the roofline to act as firewalls and perhaps to prevent burglars from climbing to the neighboring roofs (Knapp, 2005a; Liang, 1998; Liu, 1989).

Commonly, Beijing *siheyuan* had grey-tiled pitched roofs and Suzhou houses had black-tiled pitched roofs. The roof tile colors reflected social hierarchy in feudal China: yellow was exclusively used for Imperial Palaces, green for mansions owned by officials, and grey or black for ordinary houses (Liu, 2002a).

Figure 4.36   Roof detail. Photo by Donia Zhang 1991

Figure 4.37   Roof and veranda detail. Photo by Donia Zhang 1991

Interior Space

After entering a typical Beijing *siheyuan*, one would immediately encounter the inverted 'South Hall' (倒座房) that was 'sitting south and facing north,' which was normally used by male servants and gatekeepers. The Central Hall(s) serving as the living quarters for the oldest generation and guests was placed along the north-south central axis facing south, and was the highest and most exquisitely decorated. When there was no guest, the Central Hall(s) was used as a study, or place(s) for ancestral worship, conducting daily activities, holding life-cycle rituals (birth, maturation, marriage, and death), having seasonal festivities, and so on. The Central Hall(s) symbolized family unity, continuity, and the power of family clan (Knapp, 2005a).

The East and West Wing Halls (厢房) were linked to the cardinal directions and social hierarchy; they were the quarters for the lower family members such as concubines and children, and were less decorated. Spaces for wives and unmarried female family members were placed deeper in the Northern Hall(s), far from the rooms for non-family visitors and the front rooms for male servants (Knapp, 2005a).

Thus, the best buildings in a Beijing *siheyuan* were the Central Hall(s) facing south, each with two Ear Rooms (耳房) on either side. The second best was the West Wing Hall facing east, and the least ideal were the South Hall facing north and the East Wing Hall facing west. After the single-extended-family *siheyuan* turned into multi-nuclear-family compounds, a saying became common among Beijingers: "If you have money, do not live in the north-facing or west-facing halls because they are neither warm in winter nor cool in summer"[4] (Bai, 2007; Liang, 1998, p. 223; Ma, 1999; my translation).

Social hierarchy was clearly expressed in the spatial arrangement of some extensive *siheyuan*. To accentuate horizontal hierarchy, three methods were applied: (1) making the Central Hall along the north-south axis wider than the East and West Wing Halls; (2) making the Central Hall wider and deeper than the Ear Rooms on either side; and (3) making the East and West Wing Halls wider than the Ear Rooms on either side of the Central Hall (Ma, 1999).

To emphasize vertical hierarchy, three approaches were used: (1) raising the platform of the Central Hall; (2) increasing the column height of the Central Hall; and (3) raising the roof height of the Central Hall by having bigger room depth. Some wealthier families even built 2-storey Northern Halls to highlight the vertical hierarchy (Liang, 1998; Ma, 1999). The hierarchical relationships of the halls in a Beijing *siheyuan* could be expressed as such:

1. Central Hall > Wing Halls > Central Hall Ear Rooms > Wing Halls Ear Rooms ('>' denotes bigger than);
2. Central Hall ^ Wing Halls ^ Central Hall Ear Rooms ^ Wing Halls Ear Rooms ('^' indicates higher than).

---

4   有钱不住东南房, 冬不暖来夏不凉。

Figure 4.38   Interior space of a Beijing *siheyuan*. Source: Ma, 1993, p. 54

Figure 4.39   Interior space of a Beijing *siheyuan*. Source: Ma, 1993, p. 56

Inside traditional Suzhou house compounds, doors or partitions were often removable for better ventilation, and wide corridors prevented hot summer sunlight from directly entering the halls in the residential quarters (Yu, 2007).

Along the main axis of Suzhou houses, the first hall was the lobby (门厅), the second was the Tea Room (茶厅), the third was the Central Hall (正厅), the fourth was the Inner Hall (内厅), the fifth was the Northern Hall (堂屋), and the sixth was the Inner Northern Hall (内堂屋). The Central Hall was the most important, where activities such as entertaining guests, holding banquets, rituals, or ceremonies, took place. The Flower Room and the Study Room tended to be located in the best southeast orientation where privacy and tranquility were ensured (Yu, 2007).

In a large Suzhou house, parents lived in the halls along the main axis, and children in the wing halls on the sides (unless the son was married and had his own house). The rooms for the gatekeeper, accountant, private tutor, guests, footman, and the like, were on the sides, isolated from the inner rooms on the main axis (Yu, 2007), to carefully obey the imperial rule of 'setting a distinction between men and women, letting them pass in the distance.'[5]

Architectural symbolism was the *soul* of classical Chinese architecture and it contained rich symbolic meanings (Yang, 2008). Traditional Chinese people liked to decorate their houses with poetry, calligraphy, painting, pictographs, and carvings. Through associative compounds, metaphor, and homophones, these artworks expressed auspicious meanings and aspirations (appendix, table A.2) to help uplift their minds, educating their offspring, as well as communicating Confucian ethics. Although some of these meanings might seem farfetched, nonetheless they formed beautiful and harmonious pictures through skilful compositions, resulting in unique Chinese decorative arts (Knapp, 2005a). The most commonly displayed couplet would read: "Honesty and tolerance can sustain a family, just as poetry and books can last long and spread widely"[6] (Bai, 2007; Ma, 1999; my translation), which resonates with the English phrase: "Honesty is the best policy."

Floor Levels

The vast majority of Beijing *siheyuan* were single-storey, to be close to earth *qi* for health; some large houses had 2-storey buildings in the northern end of the compound (Ma, 1999; figure 4.40). Although multi-storey buildings emerged in China as early as the Warring States (475-221 BCE) (Pheng, 2001, p. 271), they were widely adopted only in southern China (Knapp, 2005a). Traditional Suzhou houses often had a second storey because of the damp ground caused by the intermittent drizzles during plum rains in June and July in the middle and lower reaches of the Yangzi River.

---

5   男女有别, 授受不亲。
6   忠厚传家久, 诗书继世长。

Figure 4.40   Aristocrats in Beijing often built 2-storey inner/northern halls, Prince Gong's Manor. Photo by Junmin Zhang 2010

Furniture Styles and Materials

The Book of Rites says: "The seasons have their timing, the earth has its crops, materials possess special usefulness, and the crafts display ingenuity" (cited in Lo, 2005, p. 173). The Chinese therefore paid much attention to the quality of furniture in their houses. The Chinese term for furniture is *jiaju* (家具), meaning 'equipment of the house.' Furniture is also regarded as 'portable architecture,' serving as a tool for human interaction in everyday life, and making the environment lively while giving it substance and meaning (Lo, 2005, p. 173).

Traditional Chinese furniture was made of hardwood in amber or purplish red hues that showed respect for organic substance. Lo (2005, p. 164) has observed four major regional furniture styles in China: (1) Beijing style with imperial court patronage, where the most skilled craftsmen gathered; (2) Suzhou and Jiangnan style directed by literati tastes; (3) Shanxi style with highly recognizable design sustained to a great extent by rich merchant-patrons; and (4) Lingnan (in today's Guangdong province) style with ornately carved furniture made of imported hardwoods, such as *zitan* and *tielimu*, or *hongmu* (types of redwood or blackwood).

There was normally a standard set of furniture symmetrically arranged along the back wall of the Central Hall of a classical Chinese courtyard house. Facing the entryway, there would be a long table to hold the ancestral altar and tablets, images

Figure 4.41   A typical set of stone table and stools (绣墩) in a courtyard for resting, a niche hosting the Daoist iconic statue 'Grand Supreme Elder Lord' (*Taishang Laojun*), Daoist Association, Beijing. Photo by Donia Zhang 2007

of gods and goddesses, family mementos, and ceremonial supplies. Shrines were also indispensable pieces of ritual furniture in traditional Chinese homes (Knapp, 2005a). The symmetry and double-set arrangement were associated with Chinese cosmological concept of *Yin Yang* balance and harmony, as well as the aesthetic principle of *xu* (void) and *shi* (solid) (Lo, 2005, p. 187).

A core piece of furniture in a traditional Chinese house was the *Baxian* dining table (八仙桌 'The Eight Immortals Table'[7]) usually made of redwood representing domestic harmony (Knapp, 1999, pp. 148-157; 2005a, pp. 92-93), and a set of stone table and stools (绣墩) was often placed in the courtyard or garden for resting.

For wealthier families, the beds were generally built on a raised platform with four posts. Beams were set on a raised base with filled curtain walls and covered

---

7   The fabled *Baxian* are Daoist Eight Immortals, who maintain their human personalities while being blessed with supernatural powers. Having lived the best of both worlds, they are viewed as spontaneous and fun-loving figures without the worry of earthly concerns or human infirmity (Knapp, 1999, pp. 148-157; 2005a, pp. 92-93).

with a canopy and usually with a 3-bay façade. These canopy beds (or alcove beds) somewhat resembled the enclosed courtyard house itself, serving as a space within a space. At night, with the silk gauze or cotton curtain drawn, it created a warm and private enclosure (Knapp, 2005a, p. 78).

Facility Provision

For Beijing residents of the past, drinking water was fetched from the well in the *hutong* and each *hutong* had its own well, which often became the center of social interaction for women (Bai, 2007, p. 9). The kitchen was generally located in an Ear Room (*erfang*) in the east of a *siheyuan*, a direction that was believed to be favored by the Stove God. Winter heating was normally provided by coal burning in a briquette furnace. The latrine was usually situated in the less auspicious southwest or northeast corner, and a 'dig-dung' worker would come to empty it weekly (Bai, 2007; Geisler, 2000).

Figure 4.42   An ear room of a Beijing *siheyuan*. Photo by Donia Zhang 1991

Every Suzhou house had one or several water wells located in the courtyards, lightwells, kitchens, verandas, Central Halls, or underneath the stairs of study rooms; some wells were shared by two rooms (Suzhou Housing Management

Bureau, 2004). Traditional Suzhou houses usually had a kitchen at the back of the compound, but no heating system provided for winter (Yu, 2007). In the old times, there was no designated toilet room in a Suzhou house but a toilet bowl made of wood was placed in each room. There were special people who came to collect the toilet bowls in the early morning, and each household was responsible for cleaning their bowls after being returned (Lynn Xu, personal communication, 2011).

The most significant change during the Qing dynasty (1644-1911) was the combination of shops/stores with residential space in the same compound; usually the shops opened to the street and the living quarters at the back, or above in the case of southern China (Sun, 2002; Yu, 2007).

Building Materials and Construction Quality

Classical Chinese courtyard houses were mainly constructed of wood. Wood as a natural material has many advantages: it is readily available, reasonably lightweight, easy to standardize, and flexible in the span. Moreover, wood is warm to the eye and touch. When a wooden structure is well protected from moisture and vermin, and painted with wood oil, it may last for a long time. Using wood as a *Yin Yang* combination material was a very suitable choice. However, wooden structures are easily destroyed by fire. Thus, not many old wooden buildings exist in China today (Knapp, 2005a; Liang, 1998; Liu, 2002a; Lo, 2005; Qiao, 2002; Steinhardt, 2002). Additionally, the massive consumption of wood as a building material has resulted in deforestation that now exposes Beijing to harsh winds.

Traditional construction materials in China also included rammed earth, sun-dried bricks, kiln-dried bricks, bamboo, and reed that were abundant in the south. In mountainous regions, stone was widely used for foundations, lower walls, columns, and window carvings (Knapp, 2005a; Liang, 1998; Liu, 2002a; Qiao, 2002).

Classical Chinese architecture and furniture applied an interlocking 'mortise-and-tenon' timber joinery system that employed notches and inserts without using adhesives or fasteners.[8] These frame structures, when built properly, can be more resistant to earthquakes, but are vulnerable to fire and natural rotting (Knapp, 2005a; Liu, 2002a; Lo, 2005; Pheng, 2001; Qiao, 2002; Steinhardt, 2002).

Classical Chinese courtyard houses were mainly composed of three parts: earthen/stone platform/podium (base), wooden column/pillar network (body), and overhanging roof framework (head). The platform prevented moisture from penetrating into the column feet and walls; it was stable enough to support walls made of rammed earth, adobe, or fired brick above, which either supported the roof structure directly or served as a curtain wall around a timber framework (Knapp,

---

8   Archaeological findings reveal that this mortise-and-tenon joinery construction method first appeared in China during the Neolithic period (6000-2000 BCE) and the oldest examples are found at Hemudu site in Zhejiang province (Liu, 2002a, p. 12).

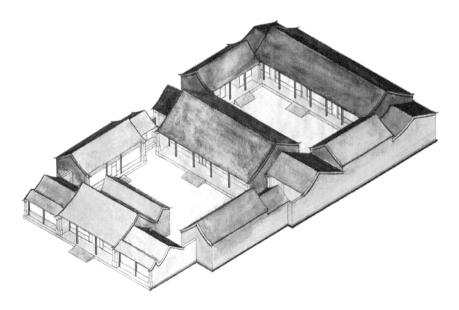

**Figure 4.43   Aerial view of part of a whole compound of a Princess Mansion
showing the platforms, columns, and overhanging roofs. The
location of the gate at the center of south wall is untypical for
Beijing *siheyuan* because the original gate was located elsewhere.
Pencil drawing and rendering by Donia Zhang 1991**

2005a; Liang, 1998). The elevated structure also provided an observation function
that made defence more readily achievable (Pheng, 2001, p. 267). Basements were
rare across China maybe because of Daoist environmental ethics as indicated in
Tai Ping Jing (The Book of Great Peace) that advises people not to dig or gouge
Mother Earth to avoid calamity (Lai, 2001, pp. 104-106).

Symbolically, in China, the platform signified Earth whereas the roof
represented Heaven (Blaser, 1995, p. 10). The salient feature of column was not
only structural, but also divided the facade into bays (*jian*) to give a rhythmic
effect. On a long building, there was normally a veranda in the front supported by
columns so that the unity between function and aesthetic could be achieved.

A standardized spatial module *jian* (间 'bay') was used for construction to
form interior and exterior spaces. This basic unit is the span between two lateral
columns/pillars. *Jian* could also be viewed as the space between four columns.
Throughout China, *jian* always had odd (*Yang*) numbers of 3, 5, 7, or 9 because
these numbers afforded symmetry and balance. The middle *jian* was normally
wider than the secondary *jian* on either side. *Jian* ranged in width between 3.3-3.6
m in northern China and 3.6-3.9 m in the south for better airflow in the hot and
humid climate. The depth of a *jian* was often 6.6 m in southern China and 4.8 m in
the north (Knapp, 2000, 2005a).

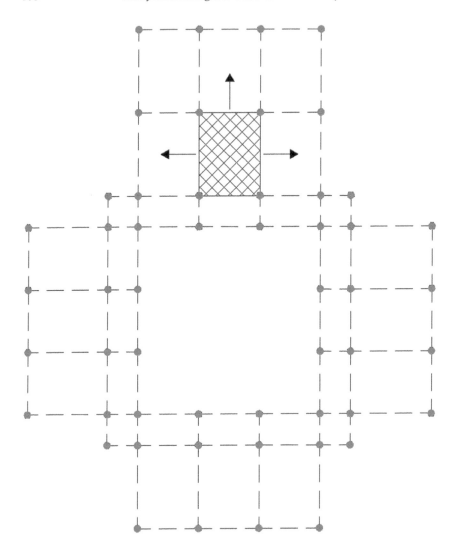

Figure 4.44 *Jian*, **a standardized spatial unit of planning and construction for classical Chinese courtyard houses. Drawing by Donia Zhang after Liu, 1989**

Social hierarchy was clearly expressed in the construction of classical Chinese courtyard houses. In the Tang (618-907) and Song (960-1279) dynasties, house construction was subject to strict regulations. The social rank of the owner determined the building height, room size, color, and roof decorations. The social

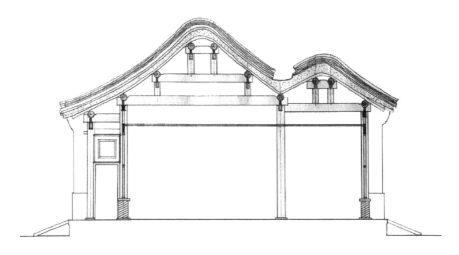

Figure 4.45   A wooden beam-column construction with 'hill ridges' (^^) roof,
               a typical classical courtyard house, Beijing. Pencil drawing by
               Donia Zhang 1991

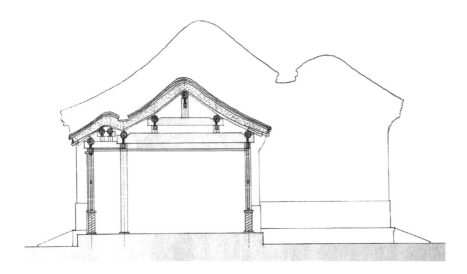

Figure 4.46   Section of an ear room and the silhouette of the front/south
               hall behind, a typical classical courtyard house, Beijing. Pencil
               drawing by Donia Zhang 1991

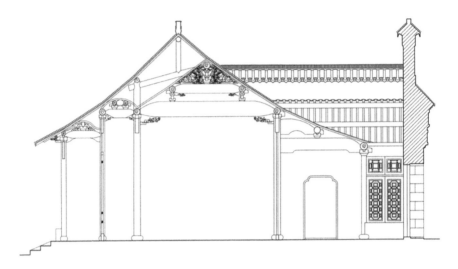

Figure 4.47   **Section and partial elevation of the Central Hall of Wu Mansion, Big Stone Lane (大石头巷), Suzhou. Source: Suzhou Housing Management Bureau, 2004, p. 97**

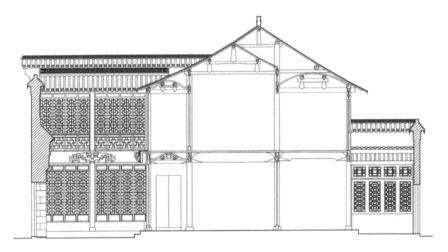

Figure 4.48   **Section and partial elevation of the Inner Hall of Wu Mansion, Big Stone Lane (大石头巷), Suzhou. Source: Suzhou Housing Management Bureau, 2004, p. 96**

hierarchy could be expressed as such: ruler ^ ministers ^ officials ^ scholars ^ businessmen ^ commoners ('^' denotes higher than). Although nobody was officially permitted to break the rule, it was not rigidly enforced and was violated throughout China. Under the less bureaucratic system of the Yuan dynasty (1271-1368), the residential restrictions were somewhat loosened (Pan, 2002; Qiao, 2002), which led to a little more creativity in house designs. Yet before 1911, no residential building was permitted to be more than one storey in Beijing's inner city so as not to exceed the height of the Forbidden City, the home of the emperor and his families. The Ming (1368-1644) and Qing (1644-1911) governments also regulated that commoners should not have a hall more than three *jian* ('bays'); if they wanted the hall larger, they could increase the depth (Knapp, 2000, 2005a).

Construction skills in ancient China were not recorded but handed down from master to apprentice or father to son, who were all anonymous craftsmen (Lo, 2005). The first Building Standards in China for the design and construction of cities, palaces, houses, and gardens was *Yingzao Fashi* (营造法式) compiled by Li Jie in 1103 (Fu, 2002b; Guo, 2002), along with *Gongbu Gongcheng Zuofa Zeli* (工部工程做法则例 'Engineering Manual for the Board of Works') issued in 1734 (Sun, 2002). Other non-official manuals included the 15th-century *Lu Ban Jing* (鲁班经, named for a 5th-century BCE craftsman *Lu Ban* who later obtained demigod status) and *Feng Shui* principles. From the early 1930s to 1966, Liang Sicheng completed the historical task of making annotations and illustrations for *Yingzao Fashi* and *Gongbu Gongcheng Zuofa Zeli*, which made them much easier to understand (Liang, 1998, p. 31).

Yuan Ye (园冶 *The Craft of Gardens*, 1631) was the first monograph dedicated to garden architecture in the world written by Ji Cheng (1582-c.1642), a Ming dynasty (1368-1644) garden designer and a Suzhou native. Another celebrated encyclopaedia on garden architecture was *Zhang Wu Zhi* (长物志 *Treatise on Superfluous Things*, 1621) written by Wen Zhenheng (1585-1645), who was also a Suzhou native and a Ming dynasty (1368-1644) scholar, painter, garden designer, and great grandson of Wen Zhengming (1470-1559, a famous Ming dynasty painter). While *Yuan Ye* focuses on gardening techniques and plants in southern China, *Zhang Wu Zhi* concentrates on the enjoyment of garden views and plants in northern China. These two volumes are not only complementary, but also touch upon some basic principles of city planning and architectural design.

Summary and Conclusion

This chapter examined the form and environmental quality, and space and construction quality of classical Chinese courtyard houses, particularly the Beijing and Suzhou types, to build the context of the study. Their common elements were the courtyard that provided natural light and ventilation, as well as natural scenery (trees, plants, flowers, water, rocks, etc.). Another common aspect was the hierarchical organization of space that reflected a Confucian code of ethics. With the harmonious integration of outstanding architecture and beautiful greenery, classical Chinese courtyard houses were works of significant quality that offered pleasant living environments. In Beijing, depending on the wealth or rank of the owner, the courtyards in more modest homes would have trees and plants, and the mansions had exquisite gardens. In Suzhou, scholars enjoyed large and lush gardens. However, service facilities were rather simple and unsophisticated, making daily life not so convenient. Not everything about a past form is a nostalgic ideal and new developments may well have alleviated some of the previous problems. A summary of the attributes of these traditional houses is presented in Table 4.1.

The chapter focused on architectural and spatial design elements in classical courtyard houses of Beijing and Suzhou. The illustrations depicted well-preserved examples containing all the important physical features. These houses were originally occupied by imperial government officials, civil servants, or scholars, and they represented different degrees of grandeur; they are comparable to the new courtyard housing being investigated because this new courtyard housing also accommodates government officials, civil servants, and scholars, among other professionals.

In Chapter 5 that follows, the social and cultural life in classical Chinese courtyard houses will be presented, to provide the context for comparison with that of the renewed/new courtyard housing discussed in Chapters 9-10.

Table 4.1   **Summary of attributes of classical courtyard houses of Beijing and Suzhou**

Characteristics	Beijing Courtyard House	Suzhou Courtyard House
1. Exterior form	Each complex 15-25 m wide and 25-50 m deep; 1-5 courtyards on north-south axis	Each complex 35-40 m deep; 5-7 courtyards/lightwells on north-south main axis
2. Exterior walls	Grey or red bricks	White plasters
3. Gate	Southeast/northwest corner	Middle of south/north wall
4. Windows	Red/green frames; best facing south, but also facing east, west, and north	Brown frames; best facing south, but also facing north; *no* east- or west-facing
5. Courtyard and gardens	40 percent of ground area, width between 7-20 m; only aristocrats had gardens	20 percent of ground area, smaller lightwells can be 1-2 sqm; scholars' gardens are outstanding legacy

Characteristics	Beijing Courtyard House	Suzhou Courtyard House
6. Roofs	Grey-tiled pitched roofs slope between 45°- 60°; eaves typically 60 cm; sometimes flat roofs for service rooms	Black-tiled pitched roofs slope between 45°- 60°; eaves typically 60 cm
7. Interior space	Hierarchical layout according to age, gender, and rank; Central Hall was the highest, largest, and best decorated	Hierarchical layout according to age, gender, and rank; Central Hall was the highest, largest, and best decorated
8. Floor levels	1-storey mostly	2-storey mostly
9. Furniture styles and materials	Hardwood/redwood in traditional Beijing style; stone table and stools in courtyard/garden	Hardwood/redwood in traditional Suzhou style; stone table and stools in courtyard/garden
10. Facility provision	Water from well in the *hutong*, kitchen in the east, heating by coal-burning, latrine in southwest or northeast corner	Water from wells in the complex, no heating in winter, kitchen at far end of the complex, toilet bowl in each room
11. Building materials and construction quality	Earthen/stone platform, wooden beam-column structures in good construction quality	Earthen/stone platform, wooden beam-column structures in good construction quality

Source: My summary

Chapter 5

Matters of Social Cohesion, Time and Cultural Activities in Classical Courtyard Houses

The riches of a family are not to be sought in its wealth,
but in the affection and harmony of its members.
Where these prevail, the family is not likely to be poor,
and whatever it has will be well preserved.

– *Yi Jing*, Appendix 2, Footnote 313:37

This is one of the three background chapters that examine the social and cultural life in classical Chinese courtyard houses to provide a context for and comparison with the renewed/new courtyard housing projects explored in Chapters 9-10. The chapter consists of seven sections: social organization and family life, relations between men and women, women in the inner quarters, cultural activities in classical courtyard houses, cultural festivities in classical courtyard houses, birthday celebrations in classical courtyard houses, and wedding ceremony in classical courtyard houses.

Social Organization and Family Life

Anthropologists who have studied China have long observed the close link between house form and family structure (Yan, 2005, p. 373). Classical Chinese courtyard houses fostered a 'harmonious hierarchy' (Li, 1993, p. 115) defined by generation, age, gender, and rank (figure 5.1) that was clearly indicated by the assignment of the halls. It emphasized the importance of an orderly family and Confucian code of filial piety towards one's parents as central to a peaceful society (Chang, 1986; Jervis, 2005, p. 228). Cosmologically, the most auspicious south-facing Central Hall along the central axis is where the ancestral altar/tablets were placed, and the bedrooms for the parents (or grandparents) and their eldest son (the ritual heir) and his wife were also located there. Younger married sons and their wives and children normally lived in the south-facing halls in an outer courtyard or further to the side of the hall (F. Bray, 2005, p. 261). Culturally, a classical courtyard house fulfilled the Chinese ideal of having several generations living under 'one roof' (Knapp, 2005a).

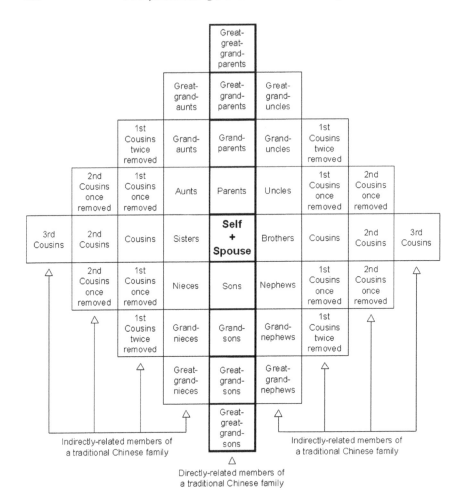

Figure 5.1 **The 'Nine Generations' family clan in imperial China. Drawing by Donia Zhang translated from Bai, 2007, p. 45**

Ancestral worship commonly took place in both northern and southern China. In the south it normally occurred in ancestral halls decorated with signs of moral admonitions that helped guide people of correct conduct (Ho, 2005, p. 322); whereas in the north it was at the grave, ancestral tablets, and ancestral scrolls (Cohen, 1990; Ho, 2005, p. 300). Through these ceremonies and rituals, those who did not live under one roof were bound together (Ho, 2005, p. 322; Faure, 2005; Jervis, 2005). Smith (1994, p. 88) noted that endorsed by Daoism, Confucianism, and Buddhism, ancestral worship strengthened the Chinese family system not only by solidifying social relationships and reinforcing status obligations, but also by maintaining a deeply conservative precedent-mindedness at all levels of society.

Nancy Jervis (2005, pp. 223-224) observes that most Chinese families belong to one of three main types: conjugal, stem, or joint. Conjugal (or nuclear) families are composed of a husband and wife and their children. Stem families was the most commonly found residential arrangement in imperial China that usually included the husband's father and/or mother. In anthropological terms, this is called a patrilineal residence pattern. Joint (or extended) families included two or more fraternally related conjugal families in the same generation, along with other members such as parents and/or unmarried siblings. However, in his book *House United, House Divided* (1976), Myron L. Cohen argues that Chinese families should not be classified as conjugal, stem, or joint; rather, each type should be viewed as being a different stage in a process that extends over time.

Joint families represent the Chinese ideal and have been celebrated in several novels such as *The Dream of Red Mansions* (红楼梦, or *The Story of the Stone*) by Cao Xueqin (1725-1764?), *Family* (家) by Ba Jin (1904-2005), and *Four Generations under One Roof* (四世同堂) by Lao She (1899-1966). It is this collective nature of traditional Chinese families that must be understood in order to fully comprehend classical Chinese courtyard houses and Chinese culture (Cohen, 2005, p. 238).

Figure 5.2   The Grandview Garden in *The Dream of Red Mansions* by Cao Xueqin (1725-1764?). Courtesy of Shanghai Library, color rendering by Donia Zhang 1996

Family contracts were an important economic aspect of Chinese life (Cohen, 2005). Before 1949, Chinese families functioned as corporate organizations and run as the basic unit of production and consumption in society. Family members usually pooled their incomes, which were then redistributed by the family head, often the father, mother, or eldest brother (Jervis, 2005). This capital accumulation had the advantage of establishing family enterprise and the members in conjugal, stem, or joint relationships could operate through common budgets (Faure, 2005, p. 281). This financial benefit was a consequence of Confucian patriarchal and patrilineal emphasis of having 'children and grandchildren filling up the halls' (儿孙满堂), all of whom should obey the authority of the senior male member of the family.

In the villages in Hebei province in northern China, only the wealthy families continually lived together after their sons' marriages. The poor families' sons moved out immediately after marriage although they still lived in the same village, whereas the daughters normally married out to other villages and lived with their in-laws (Junmin Zhang, personal communication, 2010).

In Xiangxi village of Guangdong province in southern China, David Faure (2005) discovered a case of a Zhang surname family in a somewhat different social organization and family structure. The house compound was made up of rows of modular units, most of which were separate dwellings that functioned as an independent 'family,' but were part of a joint family. The house implied that the whole group of relatives was not co-resident in the same house, and the architecture contained the notion of family division. Each dwelling might store and process its own grain, although the entire compound was to operate as a 'family' for shared cooking. Nonetheless, there was a segregation of men and women within the complex and it served as a rationalization for its unity (Faure, 2005, pp. 284-285). Another similar case was a Zeng surname family at Shan Ha village of Sha Tin district of Guangdong province in southern China (Faure, 2005, pp. 289-292).

Family dividing-up (*fen jia*) often occurred when the head of the household died, or in situations when the wives of brothers did not get along (Junmin Zhang, personal communication, 2010). When *fen jia* happened, all the family properties were equally divided among the brothers, and income was no longer pooled, but the female family members were not customarily involved in the process. Each conjugal unit would then become its own family. Upon receiving a share of the family property, the brother had become head of a new family, and the process of developing new stem and joint families started again. The new family would grow and congeal, and at some point decline and divide (Cohen, 1992, 1998, 2005, p. 237; Jervis, 2005, pp. 224-225).

Relations between Men and Women

A common feature of collective living in classical Chinese courtyard houses was to separate the male and female activity spaces to lessen the importance of conjugal intimacy (Yan, 2005, p. 375). It was normal to see an obvious indifference between husbands and wives because the axes of relations in a traditional Chinese family were that of father-son and mother-daughter-in-law, whereas the relationship between husband-wife played only a minor role. As such, obedience and hierarchy took precedence over intimacy in interpersonal interactions (Baker, 1979, pp. 11-25; Fei, 1947/1992, pp. 84-86; Yan, 2005, pp. 375-376).

As early as the 5th century BCE, Confucius had assigned women to the 'inside' (內) and men to the 'outside' (外), a division that referred not only to spaces, but also to family affairs, and the boundaries were strictly reinforced until the end of traditional era (1911) (F. Bray, 2005). For example, *The Book of Rites* states:

> Male and female should not sit together, nor have the same stand for clothes, nor use the same towel and comb, nor let their hands touch in giving and receiving. A sister-in-law and brother-in-law do not interchange inquiries…Outside affairs should not be talked of inside the threshold [of the women's chambers], nor inside affairs outside it (translated by Raphals, 1998, p. 224; cited in F. Bray, 2005, p. 259).

Imperial Chinese law permitted a man to have one legal wife admitted to his family tree. A concubine was much inferior in status and had not been presented to the ancestors when she entered the household. Even if a man's legal wife died, a concubine could not be raised to the wife's position and was forbidden to join the family rituals. Her lower rank was reflected in the less auspicious location of her room: the east or west wing hall (F. Bray, 2005, pp. 261-263).

A married man spent the night in the room of his wife or concubine, and during the day he would work in the office, shop, or fields. A scholar not serving the State would go to his study in another part of the house during the day (F. Bray, 1997, pp 136-139; 2005, pp. 261-262). Men who lingered in the inner quarters were not considered true gentlemen. In northern China where the winters are extremely cold, no outdoor work could be conducted, everyone had to stay indoors to keep warm during the day. The inner quarters would have to be rearranged to accommodate men, who would take up one bedroom and women and children another (F. Bray, 2005, p. 262; Körner, 1959, p. 6). The males and females of the inner and outer quarters joined in the same space during rituals such as ancestral worship, birthday celebrations, children reaching their age of maturity (capping for boys at 20 and pinning up hair for girls at 15), engagements and weddings, funerals, and so on (F. Bray, 2005, pp. 262-263). Children of both genders lived with their mothers; however, according to Sima Guang (1019-1086), boys had to move out of the inner quarters at the age of 10 (Ebrey, 1991, p. 33).

Gender separation within classical Chinese courtyard houses was considered crucial to social and cosmic order (F. Bray, 2005, p. 265). A lavishly illustrated Japanese book entitled *Shinzoku Kibun* ('Qing Customs,' 1799, pp. 112ff) offers a detailed first-hand account of merchant life in the affluent Chinese cities south of the Yangzi River. It describes that in the merchant houses the double doors to the women's quarters were covered with a curtain during the day and were closed and locked at night. In ordinary houses that had only one building, male visitors were greeted in the main room or on the front veranda, and the side rooms functioned as the inner quarters. For a poor household, a hanging curtain across the kitchen door might demarcate the inner quarters, behind which the women withdrew when male guests took up the single living room (F. Bray, 2005, pp. 260-261).

Figure 5.3   Sitting on the veranda enjoying the rock collection in a courtyard of Beijing; woodcut from the *Hong Xue Yin Yuan Tu Ji* (鸿雪因缘图记, 19th century). Source: Lin, 1847/1984, computer-mended by Donia Zhang 2010

Women's role was essential and complementary to that of men, such as preparing food, cooking, looking after the children and elderly, educating the children, producing textiles and clothes to contribute to the family income (F. Bray, 1997, 2005; Gates, 1996; Mann, 1997). The most productive work the wealthy women did was embroidery, which was regarded as an important cultural product and a sign of domestic harmony. Embroidery was a skill that passed down from senior to junior female family members, and quite often the products became part

Figure 5.4   A mother teaching her son to use brush pen to practice
calligraphy in a classical courtyard house of China. Source:
Ronald Knapp (2005a, p. 70) in *Chinese Houses* by Tuttle
Publishing

Figure 5.5   A young girl doing embroidery on the balcony of a classical
courtyard house in southern China. Source: Ronald Knapp
(2005a, p. 70) in *Chinese Houses* by Tuttle Publishing

Figure 5.6   A young couple husking with bellows, model at Suzhou Calm Garden. Photo by Donia Zhang 2007

Figure 5.7   A young woman grinding, model at Suzhou Calm Garden. Photo by Donia Zhang 2007

of a girl's dowry. Other domestic activities women conducted included washing clothes, going to market, and fetching water from a well in a *hutong/xiang* ('lane'). The wife of a busy official would also be likely to run the household and family property (F. Bray, 2005, pp. 266-267, 271).

Women in the Inner Quarters

In imperial China, a girl grew up on 'borrowed time.' Because of the patrilocal residence pattern, she would ultimately marry out to live in the 'inner quarters' of her in-laws' house. The family structural change forced her to share a space where she was an outsider that posed many difficulties for her. Once she gave birth to a son, her status immediate rose to a full ritual member of the marital family since she would become the mother of a member of the paternal line, a future ancestor entitled to be worshipped by her male descendents when she dies. If she gave birth to a daughter, she would be disliked (*Across China*, China Central Television series, episode 9, 2002; F. Bray, 2005, pp. 277-279).

In late imperial China (1368-1911), even the poorest families sanctified a portion of their houses as a solitary space for the unrelated females to constrain their contacts with the male family members. Except in occasional situations, decent women were not supposed to leave the inner quarters or let anyone from outside to enter them (F. Bray, 2005, p. 259). In wealthy families, the inner quarters offered each married woman a room of her own, and the canopy bed not only served as a place for sleep, but also for women's daily activities: food preparation, weaving, embroidery, entertaining friends, and so on, which were all conducted in the bedrooms and sometimes even on the bed (F. Bray, 2005, pp. 269-270; Knapp, 2005a, p. 78).

Feminist studies have shown that in late imperial China (1368-1911), the presence of separated 'inner quarters' sometimes had positive impacts on women's culture. The spaces offered them safety, delight, and empowerment when they were physically and socially segregated from men. The inner quarters emerged as the most pleasant part of the house, suitable for family gatherings and various activities, cool in summer and well warmed with braziers in winter. The spaces actually provided opportunities and fulfillments for those women brought together by marriage to form bonds of affection and companionship (F. Bray, 2005, pp. 259, 276; Jervis, 2005, p. 227; Ko, 1992, 1994).

The book *Shinzoku Kibun* (1799, p. 122) illustrates that in the inner quarters of a merchant house, the married women lived on the ground floor and their daughters on the upper floor, accessed through a staircase. The upper rooms' small windows would be covered with blinds to protect the young girls from being glanced. At the front of the top floor there built a 'dew platform,' which was a large balcony supported on pillars with wooden or bamboo balustrades and a canopy. The dew platform was a pleasant retreat from hot summer suns and served as a living room where women could do embroidery and men of the family could chat with women over breakfast or a pipe in the evenings (F. Bray, 2005, p. 268).

In late imperial China (1368-1911), a fashion emerged for urban wealthy families to hire a private tutor to come to educate their daughters in the house. There were several activities outside the courtyard houses that some wealthy women might participate in: scenic outing, boating and picnic, drinking and poetry party with female relatives and friends, or long trip accompanying their husband for official duties in a different part of the country (F. Bray, 2005, pp. 272-275; Ko 1994).

Figure 5.8   A canopy bed with associated furniture and fittings in this late
Qing-dynasty (1644-1911) drawing. Source: Ronald Knapp
(2005a, p. 78) in *Chinese Houses* by Tuttle Publishing

Figure 5.9   Women and children at leisure in the inner quarters of a classical
Chinese courtyard house. Source: Drawing by Wu Youru, 1983,
vol.2

Figure 5.10   A dew platform (露臺) suitable for domestic activities in good weathers. Source: *Shinzoku Kibun*, 1799, pp. 132-133; cited in F. Bray, 2005, p. 268 © 2005 University of Hawaii Press. Reprinted with permission

Figure 5.11   Young girls composing poetry on the veranda of a classical courtyard house. Source: Anonymous artist

Figure 5.12   Two generations playing *weiqi* (Go game) in a Beijing *siheyuan*.
              Source: Ma, 1993, p. 39

Figure 5.13   A group of children playing in a Beijing *siheyuan*. Source: Ma,
              1993, p. i

Figure 5.14 **A girl playing musical instrument for a group of gentlemen on the veranda of a classical courtyard house, model at Suzhou Calm Garden. Photo by Donia Zhang 2007**

Cultural Activities in Classical Courtyard Houses

Family life was peacefully played out in a finely tuned classical Chinese courtyard house. An idiom vividly depicts the pleasurable lifestyle in Beijing *siheyuan*: 'canopy, fish bowl, and pomegranates; master, fat dog, and chubby maid'[1] (Bai, 2007, my translation).

Traditionally, the courtyard was a space for domestic activities. Cooking was normally conducted in courtyards in summer for reducing heat indoors. Tables and stools were placed in courtyards for study or recreation. Children would play in courtyards without adults having safety concerns. Pets, plants, and flowers were also nurtured in courtyards (Wang, 1999, p. 117; Junmin Zhang, personal communication, 2010).

In southern China, such as Suzhou, where the climate is generally warm, scholars and artists would regularly meet in the courtyard-gardens of private homes where they could actively socialize, quietly contemplate, philosophize,

---

1 天棚鱼缸石榴树, 先生肥狗胖丫头。

study, compose and read poetry, paint, play chess and games, drink tea or wine, pick herbs for medicine, make elixirs in pursuit of immortality, and the like. Many of these fashionable pastimes were practised well into the Song (960-1279), Ming (1368-1644), and Qing (1644-1911) dynasties (Wang, 2005, p. 77). Courtyard-gardens thus functioned as spiritual and material refuges and facilitated a cultured way of life.

Cultural Festivities in Classical Courtyard Houses

Cultural festivals have been an important part of Chinese culture for several millennia. As early as the Spring and Autumn period (770-476 BCE), the Chinese had determined the point of Winter Solstice by observing the movement of the sun with a sundial, and created the earliest 24 seasonal division points (appendix, table A.3), which they used to guide agricultural activities and calculated some of their festival dates. However, most traditional Chinese festivals were, and still are, scheduled according to the lunar calendar (appendix, table A.4).

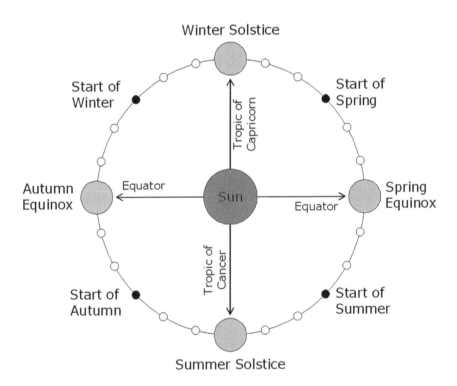

Figure 5.15   Chinese 24 solar seasons. Drawing by Donia Zhang 2009

Throughout a year, over a dozen traditional cultural festivals were celebrated at home by Han Chinese. Some festivals may have begun as agricultural events while others are associated with Chinese mythologies. Some cultural festivities required a courtyard to perform sacrifices properly to gods, goddesses, and ancestors as the courtyard linked directly to Heaven above, where the spirits were believed to reside. Most festivals are related to seasonal change, and the courtyard provided people a space to perform 'cosmic dance' by following the rhythm of nature, where one could see, feel, smell, or touch the changes of the seasons through the variations in light, air, temperature, greenery, and so on (Goh, 2004; Lee, 2004; Stepanchuk and Wong, 1991, p. ix). At a time when material goods were much less available, the abundance of food and new clothes that the festivals brought with them added to their importance in the annual cycle.

Spring Celebrations

(1) Spring Festival (or Lunar New Year) has been the most important Chinese festival, with celebrations lasting for 15 days from the 1st day of the 1st lunar month.

From the midday of New Year's Eve until the 5th day, incense, candles, and lights would be lit in houses uninterruptedly. The time between New Year's Eve to the morning of New Year's Day was believed to be the time for all the gods to descend and inspect the good and evil of the human world. People thus burned incense in courtyards to honor Heaven and Earth, and to greet every god to ensure continued blessings. The most important ritual was the sacrifice to ancestors in courtyards, Central Halls, or ancestral halls, to link past and present lineage (Stepanchuk and Wong, 1991, p. 17).

Throughout New Year's Eve, all family members would gather together to enjoy varied entertaining activities while all rooms were lit up by red candles and lanterns. Children would start singing and dancing in the middle of the courtyard at dusk to ring out the Old Year and to ring in the New Year, while adults watched them on the veranda (Lee, 2004; Suzan Li, personal communication, 2010).

On New Year's Eve, a family reunion dinner would be held in the Central Hall, where foods such as dumplings, New Year pudding (*niangao*, meaning 'promotion year after year,' usually made of glutinous rice and red bean paste), and steamed buns topped with red jujubes, would be served (Goh, 2004, p. 5; Stepanchuk and Wong, 1991, p. 21). After dinner, people would normally stay up all night playing games until dawn to greet the New Year (Goh, 2004, p. 5; Jin, 2005, p. 12). Letting off fireworks and/or firecrackers at midnight on New Year's Eve was one of the most popular activates in courtyards. Firecrackers were also let off when people offered sacrifices to their ancestors on New Year's Day, to show sincerity to follow their virtue and resume their excellence in moral behavior (Goh, 2004, p. 6; Jin, 2005, p. 13; Lee, 2004). Thus, the courtyard was central to the Spring Festivity that took place at home.

**Figure 5.16   At 12 o'clock midnight on New Year's Eve, fireworks were let off
in the courtyards of Beijing *siheyuan*. Photo by Junmin Zhang
1988**

(2) Lantern Festival (or 'Spring Spirit Festival' 上元节) is on the 15th night of
the 1st lunar month. The main activities included displaying lanterns, guessing
riddles, and eating *yuanxiao* (glutinous rice balls filled with sweet fillings)
(Goh, 2004, p. 71). Lantern display was a spectacular event originated in the
Western Han dynasty (206 BCE-23 CE) to commemorate the *Taiyi* (太一) god
who was supposed to govern all (He, 2005a, p. 22). Lanterns were often hung in
the Central Halls and verandas around courtyards from the dusk to the dawn of the
16th day (Stepanchuk and Wong, 1991, p. 35; Wang, 1999, p. 121).

(3) Blue Dragon Festival (or 'Dragon Raising its Head,' 'Blossom-tide Festival,'
'Spring-walking Festival') falls on the 2nd day of the 2nd lunar month. Its origin
is deeply related to one of the 24 solar seasons – 'Waking of Insects,' as around
this day all insects will come out causing diseases, people counted on auspicious
dragons to make plentiful rains to kill pests and facilitate spring plowing. On
the day, people would clear away injurious insects in their courtyard houses by
smoking them out (Wang, 1999, p. 121).

(4) Qing Ming Festival ('Clear Brightness Day' or 'Tomb-sweeping Day') is one
of the 24 solar seasons that falls on April 5 (April 4 on leap years), a time best
for spring plowing, sowing, planting in courtyards/gardens, as well as having an

Figure 5.17 The most unique is the 'walking-horse lantern' (走马灯) consisting of two or more wireframes, one within the other, arranged on the principle of a smoke-jack so that the air current sets them revolving. Photo collage by Donia Zhang 2010

outing. Nevertheless, the festival is more of a day of commemorating the dead and consolidating family ties. On the day, people would offer foods, flowers, and favorites of the deceased, burn incense and paper 'money' in courtyards at night, and bow before the memorial tablet in the Central Halls. To take advantage of the favorable spring winds, people would also swing and fly kites to send away worries or adversities (Bai, 2005a, pp. 30-31; Goh, 2004, pp. 87-94; Stepanchuk and Wong, 1991, pp. 61-69).

Summer Celebrations

*(5) **Dragon Boat Festival** is on the 5th day of the 5th lunar month, usually in June in Gregorian calendar. Originated in China over two millennia ago, this festival corresponds to the period around Summer Solstice (June 20-21) and is rooted in fertility rites to ensure bountiful rainfall. Prior to this midpoint of a year, nature grows (*Yang* force) to reach its highest on the day, then declines (*Yin* force) until the Winter Solstice (Stepanchuk and Wong, 1991, p. 41). Thus, Chinese people considered it especially inauspicious and did not celebrate Summer Solstice (Knapp, 2005a).

In China, it was traditionally believed that this was a time when the five poisonous creatures, snake, centipede, scorpion, gecko, and spider, reproduced. Actually, this time of a year brings hot and humid weather in southern China when mosquitoes, flies, poisonous spiders, and insects become active (Knapp, 2005a, pp. 134-135; Stepanchuk and Wong, 1991, pp. 41, 48). Before sunrise, people would go to mountains to collect mugwort (*Artemisia argyi*) leaves and calamus and hang them on doorframes upon return to avoid insects (He, 2005b, p. 37; Kou, 2005, p. 38; Stepanchuk and Wong, 1991, p. 48).

Today, the most widely known legend of the festival is in commemoration of *Qu Yuan* (c.340-278 BCE), a poet, scholar, and minister of the State of Chu in the Warring States period (475-221 BCE) who drowned himself in the Miluo River for his country. On the day of his death, a fisherman threw *zongzi* (a pyramid-shaped dumpling made of glutinous rice wrapped in bamboo or reed leaves) and eggs into the river, hoping that after eating these foods, the fish would not eat his corpse. It is said to be the origin of making and eating *zongzi* in courtyards on the day (Goh, 2004, pp. 95-104; He, 2005b, pp. 36-37; Stepanchuk and Wong, 1991, pp. 45-47).

*(6) **Bathing and Basking Festival** is on the 6th day of the 6th lunar month. As the weather becomes very hot and wet at this time of a year, everything gets rotten easily. If it happened to be a sunny day, all imperial archives, records, canons, files and documents, collected works, clothing, sheets and bedding, and so on, would be put out to dry in the courtyard, to prevent moisture from penetrating through and insects eating the books. As such, this day was also called 'Basking Clothes Festival' or 'Basking Scriptures Festival' (Chinese Online Encyclopedia, 2010), and a courtyard was essential to it. The prominent Chinese calligrapher Wang Xizhi (303-361) even had a Basking Books Platform built in his residence at Linyi, Shandong.

*(7) **Night of Sevens** (or 'Chinese Valentine's Day') is on the 7th day of the 7th lunar month. The festival originated from the Han dynasty (206 BCE-220 CE) and is related to a Chinese mythology about a young cowherd Niulang (the star Altair) and a weaver girl Zhinü (the star Vega) who are loyal to their love, a Confucian ethic advocated in imperial China. On the Night of Sevens, a festoon would be placed in a courtyard, and single or newly married women would make offerings

to Niulang and Zhinü consisting of fruit, flowers, tea, and facial powder, after which half of the facial powder would be thrown on the roof and the other half divided among the young women. It was believed that by doing so the women would be bound in beauty with Zhinü. In Beijing and Suzhou, girls would also set a basin of water in a courtyard or dew platform, and throw a sewing needle into it. If the shadow cast was thin like a thread, it was thought to be an indication of the girl being good at embroidery, a much valued skill for increasing family income in imperial China. If the shadow was thick like a stick, she would be considered unskilled (F. Bray, 2005, pp. 268-269; Stepanchuk and Wong, 1991, p. 83).

(8) Summer Lantern Festival (or 'Mid-Summer Spirit Festival' 中元节, 'Hungry Ghost Festival') is on the 15th night of the 7th lunar month. The festival derived from Daoism to commemorate ancestors and is recorded in the Daoist Canon (*Daozang*).[2] It also has a connection with Buddhism (related to legendary Urabonne or Urabanna) that evolved into a family reunion holiday, during which people would return to ancestral places when the spirits of the deceased were said to revisit the household altars. At night, people would make sacrifices by burning paper 'money' and providing offerings of gifts, foods, incense, and paper clothes in courtyards so that the spirits of the dead would not trouble the living. Children would run around courtyards with raised lotus flower lanterns in their hands, singing "lotus lantern, lotus lantern, I buy you today and throw you tomorrow"[3] (Bai, 2005c, pp. 55-57; Goh, 2004, pp. 127-132; Stepanchuk and Wong, 1991, p. 72; Wang, 1999, p. 122; my translation).

Autumn Celebrations

(9) Mid-Autumn Festival (or 'Moon Festival') is a popular harvest festival, second only to Spring Festival in China. The festival falls on the 15th day of the 8th lunar month, in the exact midpoint of the autumn season (the 7th, 8th, and 9th lunar months), and around Autumn Equinox (September 22-23) in Gregorian calendar, when the sky is usually clear and the weather fine because the moon's orbit is at its lowest angle to the horizon, making it appear the brightest and largest of the year (Bai, 2005b, pp. 42-43; Goh, 2004, p. 133; Stepanchuk and Wong, 1991, p. 51).

The custom of worshipping the moon in China can be traced back to the Xia and Shang dynasties (2000-1066 BCE). It was traditionally an outdoor activity and the food was moon cakes offered to the moon. Only when the reunion among patriarchal and patrilineal clans was later established, did the festival signify family union (Bai, 2005b, pp. 42-43; Stepanchuk and Wong, 1991, p. 51).

In China, people believed that the moon was endowed with the power of deciding human marital destiny; thus, *Yue-xia-laoer* (meaning 'the old man under the moon')

2　《道藏》："中元之日，地官勾搜选众人，分别善恶⋯于其日夜讲诵是经，十方大圣，齐咏灵篇。囚徒饿鬼，当时解脱。"

3　莲花灯，莲花灯，今天买了，明天扔。

was a matchmaker. In some places, unmarried women would light up candles and burn incense under the moonlight on this night to pray for a happy marriage (Bai, 2005b, p. 44; Stepanchuk and Wong, 1991, p. 54). Married women must return to their in-law's homes on this night if they were staying at their parents'.

Generally, Mid-Autumn festivity would start at around 8 pm when the moon is visible on a clear night sky. Family members would sit around a stone table in the courtyard, veranda, or dew platform under moonlight, worshipping the Moon Goddess, offering and eating moon cakes and fresh fruits. Poured teacups would be placed on the table, the family would wait for the perfect moment when the reflection of the full moon appeared in the center of their cups; meanwhile, they would appreciate the full moon in the dark blue night sky and enjoy the cool autumn air. Thus, a courtyard was central to this festivity. For those who were far away from home, they would relish the memory of their family members when gazing at the full moon (F. Bray, 2005, p. 268; Stepanchuk and Wong, 1991, pp. 51-55). There was a custom for Suzhou women to walk across three bridges on a moon-lit night of the festival so that they may be free from diseases and live a healthy life (*Explore*, September 2007, p. 79).

(10) Double Ninth Festival (or 'Double Yang Festival') is on the 9th day of the 9th lunar month (usually falls in October in Gregorian calendar). In *Yi Jing* (The Book of Changes), the number 6 was thought to be of *Yin* quality whereas the number 9 of *Yang* character. In standard Chinese, the number 9 is pronounced *jiǔ*, the same as for 'long-lasting' (久*jiǔ*). Thus, the number 9 in both month and day was thought to be auspicious (He, 2005c, p. 48; Stepanchuk and Wong, 1991, p. 89).

China grows a variety of chrysanthemum species, and this festival time is when chrysanthemum blossoms while most other flowers have withered, and admiring chrysanthemum blossoms in courtyards/gardens was a popular activity. Men would drink chrysanthemum wine brewed the previous year because the wine was said to have healthy effects on sharpening eyes, alleviating headaches, lowering hypertension, losing weight, and eliminating stomach pain, thus contributing to longevity. Women would put a chrysanthemum flower in their hairs or hang its branches on window- or door-frames to expel 'evil' (Goh, 2004, p. 149; He, 2005c, pp. 48, 50; Stepanchuk and Wong, 1991, p. 90).

(11) Water Lantern Festival (or 'Autumn Spirit Festival' 下元节) is on the 15th night of the 10th lunar month. Daoism suggests that there are three authorities acting on humans: Heavenly Authority, Earthly Authority, and Water Authority. Heavenly Authority grants good fortune, Earthly Authority remits punishments, and Water Authority resolves misfortune. Their birthdays fall on the 15th night of the 1st (Spring Spirit Festival/Lantern Festival), 7th (Mid-Summer Spirit Festival/ Summer Lantern Festival), and 10th lunar months (Autumn Spirit Festival/Water Lantern Festival), respectively (Bai, 2005c, p. 55).

On the Water Lantern Festival, Daoists would stick a post at their gate, on which hung a yellow flag that read: "Heavenly, Earthly, and Water Authorities, provide good weather for the crops, make the country peaceful and people safe, eliminate natural disasters, and bring good fortune." At dusk, they would raise 'heavenly lanterns' on high posts in courtyards, and make dumplings in the kitchen to present to the three authorities. They would also set flower-shaped lanterns adrift in a pond/river, and offer sacrifices to the deceased whose spirits might return for a visit on the night.

Winter Celebrations

(12) Winter Solstice Festival (or 'Midwinter Festival') is on December 21-22 in Gregorian calendar when the northern hemisphere has the shortest day and longest night. The celebration can be attributed to the Chinese belief in *Yin Yang* that, although Winter Solstice is when *Yin* qualities of darkness at its most powerful, it is also the turning point to give way to light of *Yang* attributes (Goh, 2004, p. 151). Thus, Chinese people saw Winter Solstice as a time for optimism.

Winter Solstice became a festival in the Han dynasty (206 BCE-220 CE) and thrived in the Tang (618-907) and Song (960-1279) times when it was a day for people to offer sacrifices to Heaven at an outdoor altar (Fu, 2002b, p. 98; Pan, 2002, pp. 212, 220, 222). Later, the custom evolved into a counting activity at home.

On the day, some northern Chinese would create a 'Nine-Nines Cold Dispersal Chart' in diverse styles. The Nine-Nines is a succession of nine nine-day periods between the Winter Solstice and the spring season. The first 'Nine' starts at the Winter Solstice and completes 81 days later, during which period are the coldest days of a year in northern China (Bickford, 2005; Knapp, 2005a). A courtyard would help one observe the changes of sunlight and natural scenery during this cycle.

(13) Laba Festival (or 'Congee Festival') is on the 8th day of the 12th lunar month when people would cook and eat *Laba* congee at home. The congee ingredients varied from region to region. The Beijing-style congee often included a small portion of each of the following materials: red, green, and black beans, black and white glutinous rice, dried jujubes and *longan* pulp, fruit of Chinese wolfberry, walnuts, ginkgo, lotus seeds, and plenty of water. These elements contain rich nutrition, are easy to be absorbed by the human body, and are good for health (Suzan Li, personal communication, 2009). A courtyard-garden may help grow some of these items.

From Laba until New Year's Eve, people would often affix a pair of pictorial representations of door gods in the names of Sun God and Moon God, on their double-leaf front gates to prevent 'evil' from entering their homes. There was also a variety of door god images, and the most popular were depictions of widely celebrated heroes from Chinese history and mythology. The prints for the interior doors were generally smaller than those for the front gates and integrated with cultural heroes, whose themes would often appear in perfect harmony with those on the front gates (Flath, 2005, p. 338, 341; Knapp, 1999, 2005a).

Figure 5.18
The 12 wood-relief boards hung in
the Seasonal Festive Corridor of
Suzhou Calm Garden (定园) depicting
12 traditional Suzhou festivals: (1)
Spring Festival, (2) Lantern Festival,
(3) Flower Festival (2nd lunar month,
date varies in regions), (4) Qing Ming
Festival, (5) Immortals Festival (8th
day of the 4th lunar month), (6) Start
of Summer (May 5-7 in Gregorian
Calendar), (7) Dragon Boat Festival,
(8) Night of Sevens, (9) Mid-Autumn
Festival, (10) Double Ninth Festival,
(11) Winter Solstice Festival, and (12)
Laba Festival. According to Calm
Garden's website, the corridor and the
boards act as reminders of traditional
Chinese culture. Photo collage by Donia
Zhang 2007

Figure 5.19 **The 12 stone engravings outside the Seasonal Festive Corridor of Suzhou Calm Garden depicting 12 traditional Suzhou festivals. Photo collage by Donia Zhang 2007**

(14) Thanksgiving to the Stove God is on the 23rd day of the 12th lunar month. It was a very important activity in courtyards after the lights were lit up because the stove was considered the *soul* of a Chinese house (Stepanchuk and Wong, 1991, p. 6). Conventional customs instructed that if a family could afford only one picture for the Lunar New Year, that picture must be of Stove God (Ding, 1992, p. 134; Flath, 2005, p. 334). The offerings included *tanggua* and *guandongtang* (candies made of malt, maltose, and millet) so that he would 'ascend to heaven to report good things, and descend to earth to ensure tranquility' (Knapp, 1999, p. 91; 2005a, p. 80). On the 23rd day, the old image of Stove God, blackened with dust from a year in the kitchen, was burned in the courtyard. As the smoke rose, he ascended to Heaven to report to the Jade Emperor on the family members'

behaviors since his last journey a year ago (Flath, 2005; Knapp, 1999, 2005a). A courtyard would be essential to perform this ritual.

Birthday Celebrations in Classical Courtyard Houses

Celebrations of life were important rituals in imperial China, and birthdays were customarily celebrated for children under 20 and adults over 50. The first birthday was one of the most important celebrations (Wang, 1999).

When celebrating a child's first full year of life, an event called *zhuazhou* (抓周, meaning 'the year of catching') had been wide-spread since antiquity. In the morning, the parents would take the child to worship their ancestors, come back, a table would be laid, possibly in the courtyard, with objects that the child would choose from and this was used to predict his future. *Zhuazhou* is derived from *Bagua* (eight trigrams) and can be analyzed using *Yin Yang* and *wuxing* (Five Phases) theories (Sun, 2007).

Figure 5.20   **Zhuazhou activity in a courtyard, model at Suzhou Folk Custom Exhibition Center. Photo by Donia Zhang 2007**

Figure 5.21 **The character 寿 for 'longevity' with peaches on the *Baxian* table in the Central Hall of a Beijing *siheyuan*. Photo by Donia Zhang 2007**

Traditional Chinese birthdays were calculated by the person's nominal age (the 9-month pregnancy counted as a year). The age (and the last digit of it), rank, and financial situation of a person determined the extent of his/her birthday celebration. Big celebrations would include the last digit 9 (e.g. 39...79), medium celebrations the last digit 0 (e.g. 40...80), and small celebrations the last digit 1-8 (e.g. 41-48). Significant birthday celebrations would be when one reaches 60 and 80 because five cycles of 12 years making up a complete cycle of 60 years and reaching 80 was considered 'long life' worthy of a big celebration (Chinese Online Encyclopedia, 2010; Stepanchuk and Wong, 1991, pp. 26-29).

Significant birthday celebrations would be directed by a host, and a hall designated especially for it. The hall would be decorated with a large scroll with the character 寿 (*shou*) for 'longevity.' A common birthday couplet on the front wall of the Central Hall reads: "May your good fortune be as vast as the East Sea; may your longevity match that of the South Mountain"[4] (Knapp, 1999, p. 117). On the Incense Table, fresh peaches and the Eight Immortals figures would be displayed. Birthday candles had to be in red for auspiciousness. At noon, the person having the birthday would sit in the main seat, receiving congratulations

---

4   福如东海，寿比南山。

and kowtow from relatives, friends, and younger generations of the family. After the ceremony, there would be a feast to be enjoyed by all that included wine, flour peaches, and noodles. Wealthy families might hire a group of opera singers to entertain the guests in the courtyard if the weather permitted (Chinese Online Encyclopedia, 2010; Junmin Zhang, personal communication, 2010). Thus, some key birthdays were celebrated in the Central Hall, and the courtyard played a secondary role, where celebrations might extend into it.

Figure 5.22   The traditional Chinese wedding ceremony was often performed in the courtyard. Photo from Lynn Xu and Lei Zhou during their wedding ceremony 2009

Figure 5.23 **The bride and groom bow to each other for mutual respect and love. Photo from Lynn Xu and Lei Zhou during their wedding ceremony 2009**

Wedding Ceremony in Classical Courtyard Houses

In heaven, we'd be a pair of lovebirds flying wing to wing;
On earth, we'd be a pair of joint trees growing side by side.

– Chinese proverb; my translation

The celebration of love was another important ritual in imperial China, and the main part of a traditional Chinese wedding ceremony was the bride and groom 'paying homage to Heaven and Earth,' often performed in the courtyard (or Central Hall) because the courtyard offered a direct link to Heaven and Earth.

This wedding ceremony derived from the Chinese worship of Heaven and Earth as, according to *Yi Jing*, there are three great powers in the universe: Heaven, Earth, and humans. Heaven and Earth are the natural environments where humans reside, and upon which human life relies heavily. Humans constantly pray to live in a healthy natural environment that has good weather for the crops, rich soil, abundant produce, fresh air, and clean water. As such, a traditional Chinese wedding ceremony included three bows (Stepanchuk and Wong, 1991, p. 101).

The first bow was to revere Heaven and Earth for their nourishment, the second bow to the *Yue-xia-laoer* ('the old man under the moon') for his matchmaking and witness of the marriage, the third bow to the parents for their nurturing, and lastly the bride and groom bow to each other for mutual respect and love. This marital ritual dates back to antiquity and has become a widely accepted norm in imperial China (Chinese Historical and Cultural Project, 2009).

Traditionally, only after worshipping Heaven and Earth could the couple be officially declared as husband and wife. Immediately after the ceremony, the couple would be led to their bridal chamber, where the groom could finally raise the red scarf and view the bride's face for the first time (Chinese Historical and Cultural Project, 2009), because in China's past, a woman's virtues were more highly valued than her appearance.

Summary and Conclusion

This chapter explored the social and cultural life of an extended family in classical Chinese courtyard houses to provide a context for and comparison with the empirical findings in Chapters 9-10. The historical account revealed that the spatial organization of the houses reflected and facilitated a strict but harmonious hierarchy based on generation, age, gender, and rank of the stem or joint family members.

In imperial China, there was patrilocal residence and a woman customarily married out to live with her in-laws. Men and women had different roles at home, and their communication was restricted by the spatial and social segregation of the outer and inner quarters in a house compound. Women conducted most of their productive work in the inner quarters to increase the family income, as well as look after the children and elderly, educate the children, alongside other activities. The extended family was bound by blood relations all living under one roof and regulated by Confucian ethics of filial piety and having abundant offspring. As a result, this family structure was conducive to establishing and maintaining a family enterprise.

Traditional Han Chinese cultural festivals were mostly celebrated at home. Courtyards and Central Halls were important spaces for these festivities to take place, helping to maintain traditional Chinese culture. A common theme to all the Chinese festivals is the celebration of seasonal change, and a quest to commemorate spirits and ancestors because of their belief that spiritual beings will reciprocate. This worship complies with Daoist (and Buddhist) moral principles and Confucian ethics. Birthday celebrations and the wedding ceremony were sometimes conducted in courtyards when the weather permitted.

In Chapter 6 that follows, the historic preservation and modernization of classical Chinese courtyard houses will be discussed, along with a review of recent planning policies for Beijing and Suzhou, and descriptions of six case studies examined in Chapters 7-10.

PART III
New Courtyard Housing

Chapter 6
Historic Preservation and Modernization of Classical Courtyard Houses

The constant is the useful;
the useful is the passable;
the passable is the successful;
and with success, all is accomplished.
                        – Zhuangzi, c.369-286 BCE, *The Complete Works*, Chapter 2

The 20th century was a significant turning point in the evolution of classical Chinese courtyard houses. This chapter provides an overview of this transition and evaluates some of its causes. It also introduces six case studies of renewed/ new courtyard housing projects in Beijing and Suzhou to offer a basis for the empirical findings in Chapters 7-10. The chapter consists of four sections: changes in family structure, changes in house ownership, conservation of courtyard form, and construction of new courtyard housing.

Changes in Family Structure

Although China's population more than doubled (2.3 times) between 1953 and 2010 (Census 1953; Census 2010), the family structure has decreased from extended to nuclear families, a trend echoed elsewhere in the world (Amato, 2008; UN, 2002?; Van Elzen, 2010, p. 111). Statistics show that until recently, the average household size in China had remained relatively constant at about 5.2 persons (Jervis, 2005, p. 223); it reduced to 3.96 persons in the 1990 Census, 3.44 persons in the 2000 Census, and 3.1 persons in the 2010 Census. The drop is either due to the state-imposed 'One Family One Child' policy implemented since 1978, or free choices under circumstances of rapid modernization. The vertical, parent-son relationship typically found in traditional Chinese families is being replaced by the horizontal, conjugal tie as the axis of family relations in contemporary China (Yan, 2005, p. 394). Thus, Chinese family structure has evolved from a complex corporate organization to a relatively simple conjugal unit, in which family life revolves around the couple's pursuit of financial independence, privacy, and personal space (Cohen, 2005; Yan, 2005; L. Zhang, 2010).

This change in family structure demands a subsequent change in housing form, which has implications for new housing design (Cohen, 1992, 1998, 2005; Jervis, 2005). Modern housing units are frequently built with extra rooms for the future married son and his wife, and in anticipation of the later development of a stem family (Cohen, 1992; Jervis, 1992, 2005). Similarly, in the multifamily courtyard house compounds of Beijing, the grown-up children required additional rooms in the courtyards, which made the courtyards filled with impromptu extensions. This situation may be due to the changes in house ownership, among other things.

Changes in House Ownership

Housing stock in China before 1949 had exclusively private ownership, which was in compliance with Confucianism (Chang, 1991, p. 239). However, a transformation to public property ownership occurred in 1957-1958 under Mao Zedong's socialist ideology of providing a shelter for everyone (Zhang, 1997). Public records show that in 1956, Beijing had more than 230,000 private courtyard houses transformed into public ownership, totalling 3.8 sqkm, involving 6,000 homeowners. Throughout the 1950s and 1960s, private housing was gradually transferred to local governments, resulting in many nuclear working-class families moving into traditional courtyard houses, turning the originally single-extended-family courtyard houses into 'big and mixed-yard' multifamily compounds (*dazayuan*) common in many Chinese cities. My survey results indicate that 13 percent (n=40 households) of the original homeowners in the renewed traditional courtyard houses in the northwest (Xisibei) area of inner Beijing still live together with other working-class nuclear families.

Urban housing in China since the 1960s had been under a strong centralized administrative system based on state housing provision and distribution with socialist public ownership as a major characteristic (Dong, 1987). By the end of 1970s, local governments, state institutions, and enterprises had built large quantities of public housing through industrial expansion or urban renewal, which often involved the demolition of traditional courtyard houses. Private housing had declined to 10-20 percent of the total housing stock in most Chinese cities (Wang, 1995). This system had caused many problems such as insufficient investment, severe housing shortages, injustice in distribution, extremely low rent, poor maintenance and management, and so on (Wang, 1995; Wang and Murie, 1999; Zhu, Huang, and Zhang, 2000, p. 7).

Throughout the 1980s, the role of the State in housing provision and related problems were much debated. Many Chinese economists consider that the supply of housing should be left to market forces of demand and supply, while others stress the importance of the public characteristic of housing because private developers only build for those who can afford to pay (Wang and Murie, 1999; F. Wu, 2005; Zhou and Logan, 2002). China, in its quest for modernization, has launched a series of economic reforms that let market forces and private enterprises play an increasing role in the production and consumption of housing (Nee, 1996). In 1983, the State Council assured the protection of private property rights. Many property owners of traditional courtyard houses request that the government return the ownership right to them according to the Constitution of the People's Republic of China (State Council, 1990; J. Wang, 2006; Zhou, 2006). Beijing municipal government has thus regulated to investigate the property ownership issue and return the houses to the original owners (Beijing City Planning Committee, 2002, p. 97). The State has begun a process of confirming and registering ownership titles to properties that had been held since the 1950s (State Council, 1990; Wang and Murie, 1999; F. Wu, 2005; L. Zhang, 2010; Zhou and Logan, 2002).

This policy has led to an increase in private housing investment, and consequently the State has reduced a part of its burden of housing provision. Meanwhile, various experiments have been implemented to privatize part of the existing urban public housing (Wang, 1995; Wang and Murie, 1999; Zhu, Huang, and Zhang, 2000, p. 20). In 1988, urban land leasing and housing privatization were introduced with demands that the production of housing had to be commoditized. Subsequently, real estate development was increasingly drawn into national building and urban renewal processes. Nevertheless, work units were involved in housing provision until 1998. The result was a hybrid approach to housing provision, in which the work units purchased commodity housing and sold them to their employees at discounted prices (Wang and Murie, 1999; F. Wu, 2005).

In 1992, Beijing issued a policy statement entitled Methods for Implementing the Central Government's Provisional Regulations for Leasing and Transfer of State-Owned Urban Land-Use Rights (Wang and Murie, 1999). This policy significantly stimulated the development of a real estate industry in China and uncovered the value of urban land, especially land within the inner city. High land prices have led to larger-scale renewal projects and have made them increasingly market-driven. In 1994, the Beijing municipal government delegated the granting rights of dilapidated housing renewal projects to district governments, who became the leading actors in housing renewal, real estate development, and other urban construction projects (Wu, 1999, pp. 202-205).

Since 1998, work units have stopped welfare housing provision and allowed state workers nationwide to purchase their own apartments. This transaction primarily involved the subsidized sale of work-unit owned public housing to sitting tenants because reformers argued that housing shortages were caused by the housing welfare system and the only effective way to solve the urban housing problem was to increase rents and encourage urban dwellers to own their own housing units. This housing privatization was consistent with the broader market reforms that had already taken place in China since the 1980s. Commodity housing has become a major source of housing stock in China since the late 1990s. The Chinese government has planned to use commodity housing development as a catalyst to drive urban economic development (Wang and Murie, 1999; F. Wu, 2005; Zhou and Logan, 2002).

Commodity housing in Chinese real estate markets has created a significant financial constraint on affordability; housing prices are simply outside the range affordable to most urban workers without substantial state subsidies (Zhou and Logan, 2002, p. 150). In an entirely private housing market, only a small number of urban residents could afford housing at the construction costs. The rent required to recover fully the costs of investment in new housing could exceed 70 percent of the average monthly household income (World Bank, 1992). Thus, the actual pricing scheme for commodity housing is complex. It can be sold to work units or directly to urban workers at either the government-discounted standard price or the full standard price. Housing built or purchased by work units is usually sold to workers at further discounted prices; however, the purchaser owns only a proportion of the housing unit equivalent to the proportion of the full price they have paid, the remaining proportion of the property belongs to the work unit (D. Bray, 2005, p. 174; Zhou and Logan, 2002, p. 148).

Housing prices in China further increased after commercialization in 1998, making it almost impossible for ordinary citizens to purchase a home with their wages. Meanwhile, 340,000 households in inner Beijing were required to leave their traditional homes in 2002-2007 as the authorities cleaned up the city for the 2008 Olympic Games (*International Herald Tribune,* April 23, 2002; King, 2004). Relocation-related disputes since 1995 have become a serious social and political issue (Wu, 1999, pp. 202-205).

Conservation of Courtyard Form

In the early 1950s, the prominent Chinese architect Liang Sicheng asserted that Beijing had to modernize; meanwhile, its overall cultural characteristics had to be preserved; consideration and benefit had to be given to both the old and the new (J. Wang, 2003, pp. 110, 137, 335). However, there is a difference between conserving vernacular architecture and cultural relics. The former is a functional organism, and the latter, an object. An object can be displayed in a museum for future viewing, but a living organism must serve its daily functions and cannot be left untouched if it stops fulfilling its purpose (Wang, 2004). The residents living in dilapidated courtyard houses eagerly want to improve their living conditions, but people from different educational and occupational backgrounds and age groups have different views on how to proceed with the improvement. The State Council explicitly asserts that the government should play a leading role in resolving these issues.

Historic preservation of Beijing *siheyuan* started taking steps in 1984 when the Beijing municipal government initiated a protection plan that covered only a small portion of *siheyuan* in the old city (Abramson, 2001). In 1992, the State Council approved the *Beijing Master Plan 1991-2010* that established the status of Beijing as an aspiring international city and highlighted the need for a balance between development and conservation of cultural heritage (*Beijing City Planning Chart,* 2007, p. 261).

In 1995, Beijing City and District Planning Control was launched, which restricted building height by rings of layers: the closer the new building to the city center, the lower it should be (so that the new building will not be higher than the Forbidden City).

In 2002, the *Conservation Plan of Historic and Cultural City of Beijing* was initiated that provided more detailed guidelines for the conservation of traditional courtyard houses and designated 25 patches protected status. Nine articles deserve special attention in this study:

1. Conservation and renewal of dilapidated courtyard houses must take the courtyard as the basic unit; damage should not be made to the original courtyard system and *hutong* structure (article 5.2.4);
2. The form and color of new buildings in the old city must comply with the overall style and features of the old city (article 15.0.1);
3. New multi-storey housing in the old city must be built with pitched roofs; existing multi-storey housing that has flat roofs must be changed to pitched roofs (article 15.0.2);
4. For the housing with pitched roofs in the old city, the color of the roof tiles must comply with traditional grey tone; it is prohibited to apply glazed tile roofs indiscriminately (article 15.0.3);
5. For dilapidated courtyard houses in the conservation zones, the renewal must gradually restore their traditional style and features (article 17.0.2);
6. Land development in the old city should be combined with conservation and renewal in the imperial city, and reduce population density in the conservation zones (article 9.3.3);
7. Granting permission for constructing 3-storey or higher buildings that are in disharmony with traditional style and features of the imperial city should be stopped (article 9.3.4);
8. The height of new buildings in the conservation zones must be restricted according to the height control, and inserting high-rise buildings in-between should be strictly prohibited (article 13.0.2);
9. Feasible measures and adjustments (including financial methods) to restrict the overuse of private cars in the old city should be adopted (article 11.1.4).

Thus, conserving the courtyard structure and traditional housing style, and building new housing compatible with the old, were politically endorsed. However, in China, it appears that ways are frequently found to avoid the regulations. Because 6-storey buildings are the most economical to construct, developers consistently press to be allowed to build that high at least (Abramson, 2007, p. 145; Van Elzen, 2010, p. 149).

Conserving the cultural environment must go hand-in-hand with protecting the natural environment because conservation is an expensive undertaking; it makes no sense to preserve entire neighborhoods only for them to deteriorate as a result of persistent air and water pollution (Together Foundation and UNCHS, 2002). In 1998, *Beijing's Valuable Old Trees Protection and Maintenance Regulation* was issued, which stipulates the protection of old and famous trees to increase green spaces, and transmit the city's tradition of using trees as a foil to serve buildings because old trees are organic cultural relics and are an important part of historic and cultural cities; they fit into the sustainable development strategy and provide long-term benefit to a city (Lyle, 1994, pp. 102-105).

In 2004, the State Council approved a revised *Beijing Master Plan 2004-2020*, which demands gradually changing the present single-city-center layout, and dispersing some of the functions in the old city to the new satellite towns (Wang, 2007). The revised plan designates another eight patches as conservation zones, making it a total of 33 patches in the inner city (and 10 patches in the outer city), which occupies an area of 1,807 hectares, about 29 percent of the old city. The revised plan instructs stopping large-scale demolition and reconstruction and implementing small-scale, gradual, organic renewal (*Beijing City Planning Chart*, 2007, pp. 261-266). Nevertheless, the *2004-2020 Plan* largely repeats the *1991-2010 Plan*, except with a review of the problems encountered in the 1990s, including the loss of urban fabric as a whole. Two sections offer considerably new regulations: 'Control of Building Height in the Old City' and 'Conservation of the Checkerboard Street System and the Hutong Structure' (Abramson, 2007, p. 152).

To encourage greater user participation in the conservation and restoration of Beijing *siheyuan*, in April 2004, the Beijing municipality launched *Circular Encouraging Groups and Individuals to buy siheyuan in Beijing's Old Districts and Historic and Cultural Conservation Areas*. Under this scheme, the buyers actually own the real estate and have the legal right to sell, lease, mortgage, gift, or transfer. There were about 7,000 to 9,000 *siheyuan* for sale. Majority of the buyers came from Hong Kong and Taiwan, some were returned overseas Chinese, and some foreigners (McDonald, 2004). According to a real estate agency that began dealing with *siheyuan* transactions since 2004, some people bought *siheyuan* out of personal interest, but most buyers were non-Beijingers with a solid investment in mind. In the case of *siheyuan*, having a high cultural value, the profit ratio is from 100 to 200 percent, or higher. However, there is a common concern among the buyers about the long-term stability of the *siheyuan* because the Beijing Municipal Commission of Urban Planning is not committed to the term of these conservation areas, thus the old *siheyuan* may still be replaced by high-rise buildings (Yuan, 2005).

The Suzhou Master Plan 1996-2010 states that the city has 'one body two wings.' The body refers to the old city and two wings are two new satellite towns: the New District (developed in 1990) to the west, and the Industrial Park District (developed in 1994) to the east (Zhu, Huang, and Zhang, 2000). To preserve the old city, the Suzhou municipal government and Urban Planning Bureau have likewise established housing redevelopment principles as follows:

1. Maintain Suzhou's traditional planning and design such as white walls and black-tiled pitched roofs; building height should be mainly 2-3-storeys, some parts can be 3 ½ storeys to give a sense of level change; meanwhile, improve the quality of life;
2. Make maximum use of loft spaces and underground spaces to decrease building density, increase plot ratio, and increase green spaces;

3. Construct public buildings to accommodate Residents' Committees, senior's activity rooms, and kindergartens;

4. Maximize the use of existing conditions of a site such as the original street/lane network; conserve and renovate old houses that are in good condition; preserve old trees and old wells on site; and demolish and rebuild dilapidated houses;

5. Divide roads into three types: *xiang* (lane) 7 m, *nong* (alley) 3.5 m, and sub-alley no less than 2 m;

6. Make the ratio between building height to building distance no less than 1:1;

7. Reach a financial balance between investment and profit, no more free housing; for those who do not purchase new housing but demand house ownership, they will be relocated to the New District;

8. Offer to the households social, environmental, and economic benefits as compensation for the use of their property for real estate development;

9. Concentrate bicycle parking, provide semi-basement garage;

10. Plant evergreens and fragrant flowers all year round with bushes and arbors to create changes in level; tetrastigma should be planted inside estate walls to make gable walls green (Suzhou Old City Construction Office, 1991; Suzhou Tongfangyuan Housing Estate Redevelopment Office, 1992, 1994, 1995; Suzhou Urban Planning Committee, 1992).

Suzhou Old City Construction Office (1991, 1992) has also supplemented design and construction guidelines as follows:

1. Carefully design each housing unit and strive to perfect residential functions; create large living rooms and small bedrooms; kitchens and bathrooms should be bright; interior spaces should be staggered and intricate; design some higher-standard units to satisfy different users' demands;

2. Save land and energy and utilize advanced and appropriate building technologies and fine building materials, components, and equipments;

3. Pay close attention to the organization of a supervisory team on construction sites; carefully select contractors; set up a complete construction schedule and management system to ensure the quality of construction.

Since the late 1980s, Suzhou has restored 54 inner-city neighborhoods with phased approach by combining modernization with conservation of the city's architectural and cultural heritage, preserving the exterior appearances, and improving infrastructure and interior functions. A special advisory committee was launched, including renowned planners and architects such as I.M. Pei, Wu Liangyong, Zhou Ganchi, and others. Attention was paid to the needs of the elderly, of mothers with young children, and other disadvantaged groups, daycare centers, kindergartens, primary schools, as well as recreational and sports facilities (Together Foundation and UNCHS, 2002).

Architect Ken Yeang, known for his environmentally sensitive buildings, contends that the best architecture should be like classical courtyard houses because they incorporate architectural design with the ecological environment of the courtyard; it is energy-saving and assures the residents with comfort. When asked about the massively demolished Beijing *siheyuan*, Yeang firmly states: "Rebuild them!" (J. Wang, 2003, p. 27)

Construction of New Courtyard Housing

New courtyard housing experiments in China are still in their infant stage. Until now, there are *only two* such prototypes constructed in inner Beijing, one at Juer Hutong (1990 and 1994; see Wu, 1991a, 1991b, 1991c, 1991d, 1994, 1999), and the other at Nanchizi (2003; see Lin, 2003, 2004). In Suzhou, although the number of such projects is a little more such as Tongfangyuan (1996), Jiaanbieyuan (1998), and Shilinyuan (2000), few studies have been found on Tongfangyuan (see Chen, 2011; Gong, 1999; Jin *et al.*, 2004), but none on the others. It is sometimes difficult to generalize also because architectural designs are highly contextual and site-specific; situations pertaining to one case may not apply to another. Thus, six case studies are introduced here to illustrate the unique features of each project constructed in Beijing and Suzhou. These projects represent a sensitive approach to modernizing classical Chinese courtyard houses by incorporating contemporary facilities and meanwhile, respecting Chinese courtyard tradition. These cases provide material that can be used to assess cultural sustainability (Chapters 7-10). The difference in coverage of the case descriptions is due to the availability of information and different degrees of complexity surrounding the project, and the case order is arranged according to their numbers of floor levels rather than the years of construction.

Case 1: Renewed Traditional Courtyard Houses in Beijing (2005-ongoing)

To preserve the cultural landscape of China's capital city, since 2005, the Beijing municipal government has implemented a 25-year plan to invest ¥0.5 billion CNY ($732 million USD) annually for renewing dilapidated *siheyuan* and improving the living conditions of those households in the inner city (Yuan, 2005).

The Beijing City Planning Committee (2002) stipulates that the principle for large-scale inner-city regeneration is to restore, rebuild, and remodel the dilapidated courtyard houses to their original state. The Beijing Construction Committee issued a *Technical Guide* to state that the code is to 'repair the old to appear old' (as opposed to 'repair the old to look new'). Detailed regulations were set concerning the type of roof tiles, bricks, and timber to be used so that the result is not only to maintain the practical value of traditional courtyard houses, but also to reflect the old style of Beijing. However, one third (1/3) of the old city's

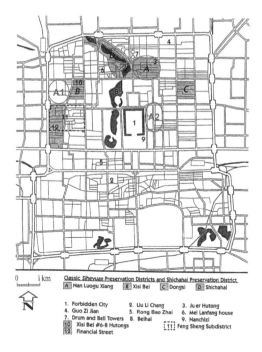

0   1 km

Classic *Sihcyuan* Preservation Districts and Shichahai Preservation District
| A | Nan Luogu Xiang | B | Xisi Bei | C | Dongsi | D | Shichahai |

⇑
N

1. Forbidden City   2. Liu Li Chang   3. Ju er Hutong
4. Guo Zi Jian   5. Rong Bao Zhai   6. Mei Lanfang house
7. Drum and Bell Towers   8. Beihai   9. Nanchizi
10 Xisi Bei #6-8 Hutongs   11 Feng Sheng Subdistrict
12 Financial Street

Figure 6.1   Map of Beijing showing the three cases' locations: renewed traditional courtyard houses (A1), Nanchizi new courtyard housing estate (A2), and Juer Hutong new courtyard housing estate (A3). Source: adapted from Abramson, 2001, p. 11

population has to be relocated elsewhere (Beijing City Planning Committee, 2002, p. 98; Kong, 2004).

The case of renewed traditional courtyard houses in the northwest of the Forbidden City is adjacent to Beijing's Xisibei conservation zone. Historically, the area was an exclusively residential district where lived high officials and dignitaries. Most of the houses were built in grand scale and style at the end of the Qing dynasty (1644-1911) and the early Republic of China (1911-1949). Although a great many have had 'wear and tear' to varied extent, their timber structures are still in good standing. About 40 percent of such buildings needed repair and restoration. The improvised extensions in each courtyard had occupied about 50 percent of the site, totalling 75 percent of the built-up areas in each compound, destroying the original forms and aesthetics of these houses, blocking light and air, and greatly reducing the quality of life (Beijing City Planning Committee, 2002).

The Beijing City Planning Committee (2002) thus instructs that making adjustments to the unplanned buildings inside each courtyard will be an effective way to conserve these traditional courtyard houses. Because conservation areas often have high population densities but few planned public green spaces, the greening of each courtyard is paramount, for example, to preserve precious, mature

Figure 6.2   **Renewed traditional courtyard house at Liuhai Hutong, Denei Street, Beijing. Note the narrow distance between the buildings. Photo by Donia Zhang 2007**

Figure 6.3   **Renewed traditional courtyard house at Qingfeng Hutong, Beijing. Note the ad hoc extension is also renewed. Photo by Donia Zhang 2007**

trees inside each courtyard (pp. 96-98). Under the premise that architectural forms and landscapes will be unaffected, 'sunken courtyards' or extra outdoor canopies are encouraged to be built for different functional requirements (Kong, 2004).

The facilities in these traditional courtyard houses were incomplete and the main problems were inadequate electric power capacity and therefore significant use of coal heating, resulting in severe air pollution. Since the area was not equipped with gas pipes, most households had to use liquefied gas tanks for cooking. Many courtyard compounds shared one tap, water supply became inconvenient for the residents, and drainage pipes were often blocked due to long periods of disrepair. However, 3870 households lived in these dilapidated courtyard houses, of which 9942 persons were permanent residents, with another 1000 floating population from outside Beijing. Workers were the largest single group of the population, accounting for 34 percent (Beijing City Planning Committee, 2002, pp. 96-97).

The Beijing City Planning Committee (2002) therefore demands that the principles of interior renovation is for buildings to fully serve contemporary living functions with increased electric power capabilities, modern bathrooms, kitchens, and water pipes (although not mentioning specific provisions for each household). Restoring Beijing's traditional *siheyuan* style and features may also mean adding a basement level to fully utilize underground space (Kong, 2004). Hence, improving the living conditions of Beijing's ordinary citizens was a major concern of the municipal government.

Case 2: Nanchizi New Courtyard Housing in Beijing (2003)

Nanchizi (南池子 'South Pond') new courtyard housing estate occupies 6.39 hectares of land in the old Imperial City in the east of the Forbidden City, around the Pudusi (Buddhist) Temple and in close proximity to the Cultural Palace, the Changpuhe Park ('Calamus River Park' or 'Iris River Park,' 2002), [1] and the Imperial Walls. The site was originally occupied by royal warehouses. After 1911, the quarter was built with ordinary single-storey courtyard houses without service facilities (Beijing City Planning Committee, 2002; Lin, 2003, 2004).

Before redevelopment, the houses' timber structures were rotten and roofs leaking, and many houses posed safety issues. The infrastructure was poor and the drainage was often blocked; only a few *hutong* were provided with running water and electricity but without a heating system or natural gas; coal-burning had caused much pollution and brought many inconveniences to the residents. Public toilets in the *hutong* were over 100 m away for many households. Of the 17 *hutong* around Nanchizi, most were less than 5 m wide and some were a dead-end, the self-built extensions further narrowed the *hutong*, making them impossible for

---

1   The park is on the east side of the Forbidden City that used to be a children's playground for the imperial families. The site was occupied by self-built shelters after 1949 and was cleared for the park, which opened in 2002.

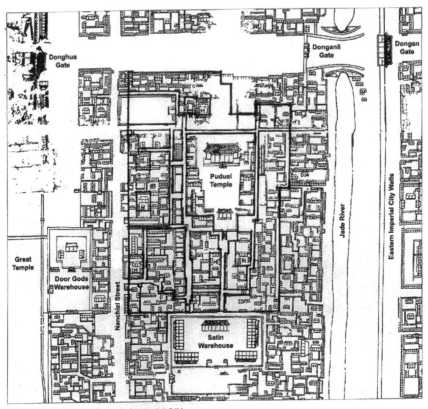

乾隆《京城全图》中本地段街巷凯图 (H 为格局规整的宅院)

Figure 6.4   **Nanchizi area map in the Qianlong period (1736-1795), Beijing. It shows a clear courtyard pattern with Pudusi Temple at the center. Source: Wang, 2004, p. 106**

fire engines or ambulances to drive through in case of emergency (Beijing City Planning Committee, 2002).

Nevertheless, 1060 households lived in 600 dilapidated courtyard compounds in the area with a density of 480 persons/ha (Lin, 2003). To achieve historic preservation for the area, a large number of the original residents had to relocate elsewhere. The new building height was restricted to be less than 6 m and to maintain the original architectural scale, form, color, materials, roof shape, and so on, of the surrounding historic buildings, and to ensure the continuity of the existing urban pattern while establishing a service network system (Beijing City Planning Committee, 2002; Lin, 2003; Shu, 2004).

The project has therefore completely restored the Pudusi Temple, preserved nine *hutong*, renovated 31 *siheyuan* that were in good condition, reconstructed 17 *siheyuan* based on their original layouts, and maintained 64 arbor trees. It has

Figure 6.5   Aerial view of Nanchizi new courtyard housing estate (2003) adjacent to the Forbidden City, Beijing. Source: Nanchizi Project Construction Headquarters, 2004, p. 98

also built 301 numbers of 2-storey Chinese-style town/terraced new courtyard houses clustered in 49 compounds (*sihelou*, meaning 'buildings with four-sided enclosure'); most of the new courtyards are shared by 4-15 families[2] (only a few single-family ones).[3] Municipal service facilities (running water, gas, electricity,

---

2   According to the project architect Lin Nan, the original design for new Nanchizi was 4-6 households sharing a courtyard (see Lin, 2003).

3   Nanchizi new courtyard housing estate occupies an area of 6.39 hectare, of which 4.3 hectare within the height restricted zone. The area is mainly residential 52 percent,

**Figure 6.6   A new courtyard at Nanchizi, Beijing. Photo by Donia Zhang
2007**

drainage, telecommunication, underground parking, etc.) have been provided
(Kong, 2004; Lin, 2003). The *hutong* were widened and some opened up for
fire engines and ambulances and for burying service pipes. Planning regulations
required to have 30 percent returning households, but only 27 percent (290/1060)
had returned, the rest had relocated elsewhere (Beijing City Planning Committee,
2002; Lin Nan, project architect, interview, 2008; Nan, 2003).

The Nanchizi renewal project was to fulfil basic living requirements of ordinary
citizens, with its spatial design based on the National Building Standards at the
time (2002-2003). With a fixed plot ratio, each household only has the minimum
area of 45-75 sqm set by the Standards and household size. Each house has its own
kitchen, bathroom, dining room, and bedroom(s), the present condition is a major
improvement over previous ones on site (Lin Nan, project architect, interview, 2008).

---

retail along Nanchizi Street 5 percent, and government offices 10 percent, building density
42 percent, and population density 240 persons/ha. The average living space has increased
from 27 sqm to 69 sqm, public green space from 1 to 6 percent, cultural and recreational
space from 0.4 to 1.4 percent, and public parking space from 0.5 to 1 percent (Beijing City
Planning Committee, 2002).

Because Nanchizi new courtyard housing was the first of its kind implemented in a historic and cultural conservation site, it has always been the focus of attention. The property value has also increased significantly. At the time of completion (2003), it was sold at ¥6,000 CNY ($885 USD) per sqm to the returning households; only the wealthier families could afford to return. In 2007, it went up to ¥15,000 CNY ($2,211 USD) per sqm. In 2008, someone from Shandong Province proposed to buy all the houses in two compounds for ¥20,000 - ¥30,000 CNY ($2,950 - $4,425 USD) per sqm (5 times of its original price), but nobody wanted to sell because of its close proximity to the Imperial Palace (interviews, 2008).

Case 3: Juer Hutong New Courtyard Housing in Beijing (1990 and 1994)

Juer Hutong (菊儿胡同 'Chrysanthemum Lane') new courtyard housing estate occupies 8.28 hectares of land (Wu, 1999, p. 114) in inner Beijing's Eastern District. It was a housing renewal experiment in a dilapidated traditional courtyard house neighborhood called *Nanluogu Xiang* ('Gong and Drum Lane South'). *Nanluogu Xiang* is a small lane near the Bell and Drum Towers recently evolved into a commercial street filled with exotic shops, bars, and restaurants, which have attracted foreigners working in Beijing to live nearby, to enjoy a lively night life (Davey, 2000).

Juer Hutong (438 m long and 6 m wide) intersects with *Nanluogu Xiang* to the east; it was formerly home to Ronglu, a governor during the Qing dynasty (1644-1911). It was a typically decayed area that urgently needed remodelling, with a terrain of 80-100 cm below the street level due to road reconstructions. The courtyards were filled with improvised extensions and two-third (2/3) of the households could not receive sunlight. Nearly 800 people lived there with an average floor space of only 7.8 sqm per person. There was one water tap in each courtyard, one sewer exit, and a public toilet 100 m away (Wu, 1991c).

Since 1978, Professor Wu Liangyong and his planning team at Tsinghua University spent a decade researching 'organic renewal' for historic cities and designed Juer Hutong new courtyard housing (or 'quasi-courtyard housing'). Phase one (4 courtyards with 46 units) was completed in 1990 and phase two (11 courtyards with 164 units) in 1994. A flexible courtyard system was adopted to fit in between the houses in good condition and those whose owners were unwilling to participate in the project. This prototype has 'borrowed' the composition principles of large mansions in Suzhou's vernacular architecture and applied it to Beijing by having 2-3-storey walk-up apartments grouped along the horizontal and vertical circulation lines, with a series of courtyards developed from the south to north and a row of courtyards from the east to west, forming a basic residential block to satisfy the demands of multi-household residence. The integration of housing with the site is thus maintained due to the compatibility between the old and the new courtyard systems in the city, meanwhile, the old trees and the *hutong* have also been preserved (Wu, 1991a, 1991b, 1991c, 1991d, 1999).

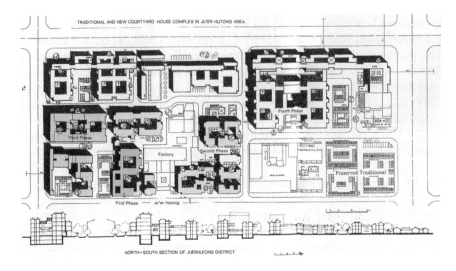

Figure 6.7 **Juer Hutong new courtyard housing estate showing the four phases and protected traditional courtyard houses, Beijing. Source: Information Center (previously Resources Center) of the School of Architecture at Tsinghua University**

The infrastructure and the physical living condition have been improved by providing each unit with privacy and spaces for utilities (kitchen, bathroom, balcony, terrace) that did not exist in traditional courtyard houses, at the same time achieving a relatively high density (928 persons/ha) and plot ratio (1:1.32) (Wu, 1991c, 1991d, 1999). The spatial design has complied with the late 1980s' Building Standards, so the apartments in phase one are generally small, ranging from 40-60 sqm, the biggest unit is 120 sqm in phase two, with more than half at 80-90 sqm. The concept of 'hall' (*ting*) was a 'cross hall' (*guo ting*) at the time, but not the 'living room' (*ke ting*) as commonly understood today (Liu Wenjie, project manager, interview, 2008).

The Juer Hutong new courtyard housing was the first of its kind built on existing traditional courtyard houses site and was an official task supported by the Beijing municipal government at the time. Its phase one has won six awards, including the *World Habitat Award* in 1992 (Wu, 1999). Its phases three and four were designed but construction was suspended due to the rising land value, the loss of government subsidies, and the developer's concern about a lack of profit[4] (Chen Zhijie, interview, 2007; Zhang and Fang, 2003). The problem lies in funding issues beyond the scope of architecture (Liu Wenjie, interview, 2008).

---

4   The Beijing municipal government requires that for a housing regeneration project, at least 1/3 of the original residents have to be re-accommodated on site, but this regulation will not be profitable enough for developers (Chen Zhijie, interview, 2007).

Figure 6.8   Aerial view of Juer Hutong ('Chrysanthemum Lane') new courtyard housing, Beijing. Source: Information Center (previously Resources Center) of the School of Architecture at Tsinghua University

Figure 6.9   A new courtyard at Juer Hutong, Beijing. Source: Information Center (previously Resources Center) of the School of Architecture at Tsinghua University

The Juer Hutong project was intended to have one-third (1/3) returning residents to maintain the original community structure (Liu Wenjie, project manager, interview, 2008; Wu, 1991c, 1999). However, when its phase one was completed (1990), it was sold at ¥600 CNY ($89 USD) per sqm to the original households and ¥3,000 CNY ($442 USD) per sqm to other purchasers. Some workplaces subsidized at ¥250 CNY ($37 USD) per sqm, and the buyers paid ¥350 CNY ($52 USD) per sqm. Some workplaces were unable to subsidize, then the ¥250 CNY per sqm was exempted, the buyers just paid ¥350 CNY per sqm to purchase their units. As such, only 10 percent of the original households could afford to move back (Chen Zhijie, interview, 2007).

Case 4: Tongfangyuan New Courtyard-Garden Housing in Suzhou (1996)

Tongfangyuan (桐芳苑 'Aleurites Cordata Fragrant Garden Estate') occupies an area of 3.61 hectares in the Pingjiang Historic District in the northeast of the old city of Suzhou, bordered by the Garden Road, White Pagoda Road East, Lindun Road, and Lion Grove Temple Alley. It is adjacent to the famous Lion Grove Garden, at walking distance from the Humble Administrator's Garden[5] in the north, the East Garden in the east, and 400 m away from the North Temple Pagoda in the west, about 900 m from the city's busiest shopping area, Guanqian Street. Before redevelopment, the low-lying site housed 600 households in small and shabby traditional houses built in the Qing dynasty (1644-1911), 75 percent were dilapidated, and the living condition was very poor.

Tongfangyuan was the first housing redevelopment in the old city of Suzhou using the local vernacular architecture concept. Supported by the Suzhou Construction Bureau, its construction started in 1992 and was completed in 1996 with a ratio of regeneration to preservation of 6:4. It has adopted traditional white walls and black-tiled roofs, as well as traditional patterns of small paths opening-up on enchanting vistas, maintaining the traditional Suzhou grid system in cross road 田 ('field') structure. The original lanes and their names are also kept such as *Tongfang Xiang* ('Aleurites Cordata Fragrant Lane'), *Saijin Xiang* ('Matching Gold Lane'), *Tianming Nong* ('Bright Sky Alley'), and *Zhangcaiyuan Nong* ('Zhang Vegetable Garden Lane'). In addition, ancient relics, residences, old trees, and wells are preserved, which include *Dujiao Ting* ('Single Corner Pavilion'), *Mengmei Ting* ('Dream of Plum Pavilion'), and *Shamao Ting* ('Gauze Cap Pavilion') (Suzhou Old City Construction Office, 1991; Suzhou Urban Planning Committee, 1994).

Tongfangyuan has been transformed into a unique residential district with complete infrastructure. There are four gates in classical Suzhou style, one in each cardinal direction. The north and south gates do not line up directly to avoid strong cross-wind and to have a view facing each gate (*Feng Shui*). Inside the estate,

---

5   Both the Lion Grove Garden and the Humble Administrator's Garden are on the UNESCO World Heritage list.

Figure 6.10   Map of Suzhou showing the three cases' locations: Tongfangyuan (A4), Shilinyuan (A5), Jiaanbieyuan (A6). Source: adapted from the web

Figure 6.11   Map of Tongfangyuan new courtyard-garden housing estate, Suzhou. The planned density is 326 persons/ha, building density 59 percent, and plot ratio 1:1.02 (Suzhou Old City Construction Office, 1991; Suzhou Urban Planning Committee, 1994). Photo by Donia Zhang 2007

public and semi-public spaces are divided by low walls pierced by moon gates, making them transitional spaces to private homes. Three types of housing forms are observed: 2-storey Chinese-style town/terraced houses, 2-storey courtyard villas, and 3-storey apartment buildings, totalling 220 housing units, each provided with modern facilities. Retail and office buildings of 4 storeys were built along the perimeter of the enclosing walls facing the street, resulting in a 'walled estate' to keep noise out. Tongfangyuan has a small, communal garden of 325 sqm located in the northwest corner, and a kindergarten at the center of the estate.

Suzhou's planning policies have been less restrictive to allow architects to design larger and more sensible living spaces, arguably in a warmer climate. The Suzhou City Planning Bureau requires housing interior space to be practical, comfortable, and aesthetic, and each apartment at Tongfangyuan to be about 90 sqm, with town/terraced houses 100 sqm, and only a few villas of 200 sqm (Qu Weizu, project planner, interview, 2008; Ren Huakun, project architect, interview, 2008). Professor Wu Liangyong was a design consultant for the project.

The project has won several awards, including the *State Pilot Award* by the Ministry of Construction in 1997, the *China Human Settlements and Environment*

Figure 6.12 Tongfangyuan new courtyard-garden housing estate, Suzhou. Photo by Donia Zhang 2007

Figure 6.13 A view of Tongfangyuan semi-public courtyard through the moon gate, Suzhou. Photo by Donia Zhang 2007

Award in 2001 (Together Foundation and UNCHS, 2002), and the First Award in Suzhou Residential Design Competition. Several other estates in Suzhou followed this design concept with some increases in interior spaces (Ren Huakun, project architect, interview, 2008).

Tongfangyuan was initiated by the Ministry of Housing and its intention was to allow a quarter (1/4) of the original residents to return (Gong, 1999). However in 1994, the sales price was ¥3,000 CNY ($445 USD) per sqm and in 1996, the price went up to ¥4,000 CNY ($587 USD) per sqm when the average market price was ¥600 - ¥700 CNY ($89 - $103 USD) per sqm in Suzhou. Only a minority of the original households could afford to return; most buyers were business people or civil servants (Ren Huakun, interview, 2008), which has resulted in an influx of high income households and has significantly altered the neighborhood structure (Y.C. Wang, 2003).

Case 5: Shilinyuan New Courtyard-Garden Housing in Suzhou (2000)

Shilinyuan (獅林苑 'Lion Grove Garden Estate') occupies an area of 3.32 hectares and is located beside Tongfangyuan; the two housing estates are separated by the Garden Road. The site was originally the Suzhou Elastic Weaving Factory with slightly over 100 old houses, and the factory was forced to shut down by the government because it drained a large amount of polluted water containing toxic waste (Suzhou Sujing Real Estate Development Corporation, 1996; Suzhou Urban Planning Bureau, 1997).

Shilinyuan (2000) was built 4 years after Tongfangyuan (1996) and so its design was somewhat improved. For better sales, the Sujing Real Estate Development Corporation (a private developer) allowed wider alleys and a bigger communal Central Garden as a communication space. The estate has 30 percent landscaping and 232 units in two forms: 3-storey Chinese-style town/terraced houses each with a small private courtyard, and 3-4-storey parallel apartment buildings enclosed by low walls pierced by moon gates on two sides, creating a series of semi-public courtyards resembling traditional lightwells. White exterior walls and black-tiled roofs are the norm, although traditional architectural vocabulary seems to be more simplified.

Like Tongfangyuan, Shilinyuan is surrounded by retail units along the estate's enclosing walls facing the street, selling arts, crafts, and souvenirs. Suzhou planning regulations strictly prohibited the building of restaurants there to minimize noise and pollution (Qu Weizu, project planner, interview, 2008).

Shilinyuan was an 'affluent' (*xiaokang*) experimental project set up by the Ministry of Housing so that the spatial design would fulfil middle-class living requirements with more innovation. The living space needed to be bigger, the functions more rational, and the overall project more original. For example, when an apartment's normal ceiling height is 2.8 m, a duplex apartment was only given 5.5 m in height so that the architect had to work out the volume to accommodate more units (Peng Hongnian, project architect, interview, 2008). After completion, the project went through rigorous examination by the Suzhou Environmental

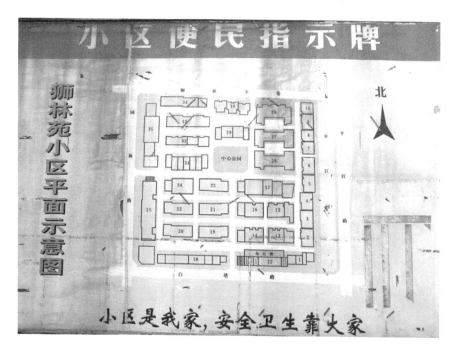

Figure 6.14   Map of Shilinyuan new courtyard-garden housing estate, Suzhou. Photo by Donia Zhang

Figure 6.15   Shilinyuan town/terraced houses, each with a small, private courtyard, Suzhou. Photo by Donia Zhang 2007

Figure 6.16   Shilinyuan apartment buildings with a semi-public courtyard, Suzhou. Photo by Donia Zhang 2007

Protection Bureau and won the *Lu Ban Award* (a national award to high-quality construction projects) by the Ministry of Housing.

Shilinyuan was sold at ¥3,200 CNY ($472 USD) per sqm at the time of completion in 2000, and the apartments on ground floors were a little cheaper. It was more expensive than the new Industrial Park District (slightly over ¥1,000 CNY [$148 USD] per sqm then) because Shilinyuan is close to the city center and only 5-6-minute walk to a hospital, which is very convenient for the elderly. Its unit price increased to ¥12,000 CNY ($1,770 USD) per sqm in 2010.

Case 6: Jiaanbieyuan New Courtyard-Garden Housing in Suzhou (1998)

Jiaanbieyuan (佳安别院 'Excellent Peace Garden Estate') was built in the Canglang District of the old city of Suzhou. The site is close to the Cultural Temple, the old campus of Suzhou High School, the Master-of-Nets Garden, and the Canglang Ting ('Surging Waves Pavilion'). Originally, there was a factory on the site with low-lying terrain that needed remodelling. After redevelopment, the estate has over 500 housing units in two types: 2-storey Chinese-style town/ terraced houses each with a small, private courtyard, and 4-6-storey apartment buildings enclosed by low walls pierced by moon or flower gates on two sides. There is a communal swimming pool, a tennis court, and a covered car park at

Figure 6.17   Map of Jiaanbieyuan new courtyard-garden housing estate, Suzhou. Photo by Donia Zhang 2007

ground level. The communal Central Garden was designed in the style of classical Suzhou gardens and is the largest and most exquisite among the three Suzhou cases. A security guard indicated that the units were sold at ¥2,800 CNY ($413 USD) per sqm at the time of completion in 1998, and the price went up to over

Figure 6.18   Jiaanbieyuan Chinese-style town/terraced housing, Suzhou. Photo by Donia Zhang 2007

Figure 6.19   Jiaanbieyuan apartment buildings, Suzhou. Photo by Donia Zhang 2007

¥10,000 CNY ($1,475 USD) per sqm in 2007. The homeowners are mostly business people, civil servants, teachers, and medical doctors.

Summary and Conclusion

This chapter provided background information on the evolution of classical Chinese courtyard houses since 1949, along with recent historic preservation policies pertaining to housing (re)development in Beijing and Suzhou. It indicated that the rapid and massive demolition of traditional courtyard houses was due to population increase, changes in family structure, and changes in house ownership, among other things, and that these have resulted in a change in dwelling culture in China.

The chapter then described six cases of renewed/new courtyard housing projects constructed in Beijing and Suzhou since the 1990s, to illustrate different approaches to preserving the courtyard form and modernizing classical Chinese courtyard houses. This research will explore whether these projects are culturally sustainable, and how they function and facilitate the continuity of cultural activities and festivities.

In Chapter 7 that follows, the form and environmental quality of the renewed/new courtyard housing will be explored, to see whether they are in harmony with heaven, and whether they satisfy residents' living requirements.

Chapter 7

Harmony with Heaven: Form and Environmental Quality of the New Courtyard Housing

They who accord with heaven are preserved,
and they who rebel against heaven perish.
— Mengzi, 372-289 BCE, *Works*, Book 4 Part 1 Chapter 7

This chapter analyzes the findings on the architectural and environmental qualities of the renewed/new courtyard housing in Beijing and Suzhou, such as the way and extent to which daylight access and natural ventilation are achieved. These factors contribute to the concept of harmony with heaven, which is one of the four 'harmonies' identified in Chinese philosophy (Chapter 3), and one of the four cornerstones of culturally sustainable architecture in China (Chapter 11). The aspects studied include: exterior form, exterior walls, gate and access, windows, courtyards and gardens, and roofs.

Exterior Form

This research shows that despite the good intentions of Beijing's conservation policies, the outcome of the renewed traditional courtyard houses (Case 1) is less than satisfactory to the residents, with numerous issues raised. For example, a resident found that completed façades do not always match the drawings submitted for planning permission. Another resident noted a discrepancy between the number of houses which applied for and were granted renewal, and the number actually renewed. Moreover, when households are randomly selected, there is neither integrated planning nor thorough renewal, resulting in a non-unified urban scene.

The Nanchizi (Case 2) new courtyard housing experiment aroused many social debates. Those who were against redevelopment mostly came from cultural circles embracing nostalgia. Many residents who lived in dilapidated courtyard houses desperately wanted redevelopment to improve their living conditions. In Nanchizi lived some leaders of the central authorities, who said before redevelopment that the area was too old and shabby, and that immediate improvement was necessary. Since the site is next to the Forbidden City, the Beijing City Planning Committee sent a delegate to UNESCO to explain and ensure that the new buildings would be in harmony with the Imperial Palace (Lin Nan, project architect, interview, 2008).

Figure 7.1   Renewed traditional courtyard houses at Suluobo Hutong, Beijing. Photo by Donia Zhang 2007

As Nanchizi area has a height restriction of 6 m to the eaves, most of the buildings have 2 storeys, with 3 storeys only in some parts that are still within the 6 m height limit to create differences in roof levels. From the developer's point of view, the key to the project's success lies in the achievement of maximum building height and plot ratio (1:1.6) within the planning regulations (head developer, interview, 2008).

When asked, "How does the *form* (such as exterior appearance, gate location, sunlight, ventilation, roof design, etc.) of the renewed/new courtyard housing help or hinder your daily/cultural activities?", residents at Nanchizi had different views. Seven of 16 interviewed residents commented positively that the form of the new courtyard housing looks fine and splendid, that it has maintained the style and features of the ancient capital and traditional *siheyuan* that turns 1 storey into 2 storeys, and that it allows people a taste of the small courtyard house's lingering charm. However, six residents argued that although the environment has been transformed, it cannot really be called a 'courtyard house' because the shape of each yard is different. They also contended that it is a new 'mixed-yard' that has destroyed the traditional *siheyuan* layout, and that when combining Chinese and Western architectural styles, the design looks 'neither fish nor fowl.'

The Juer Hutong new courtyard housing (Case 3) was designed so that the new insertion would take into account the old scale of the city to create harmony between the two. For its exterior colors, the designers used colors common

Figure 7.2   A typical compound at Nanchizi new courtyard housing, Beijing. Photo by Donia Zhang 2007

in southern China: white walls, black-tiled roofs, and light-brown gables to symbolize wood. Although traditional wall colors in Beijing are grey and red, grey would seem depressing in the small housing estate while red would look too strong. Hence, white is preferred as neutral and bright (Liu Wenjie, project manager, interview, 2008).

Regarding the exterior form, three of 17 interviewed residents at Juer Hutong commented positively that the new courtyard form is good, irregular, and attractive; they also found the structure nice, staggered, intricate, and unique. Since not all the original residents moved away at once, the design was carried out piecemeal, resulting in such 'staggered and intricate' shapes. One resident praised the design:

> The exterior appearance is good. As a first group of original residents, I was deeply attracted by its Minzhou cultural characteristics exhibited: white walls and black-tiled roofs.... Juer Hutong new courtyard housing represents a unique architectural model by combining historic Beijing *hutong* and *siheyuan* culture with southern Chinese vernacular architecture. It has straddled local cultural traditions to maintain Chinese cultural roots.

The new, classical-style street lighting at Juer Hutong was assessed by a resident:

The newly installed street lamps look classical and more elegant than the old ones, but they are too low – about 2 m high that can easily be vandalized by children [which has happened before]. It may take a long time for the city to come to replace a bulb if that happens.

Figure 7.3   Juer Hutong new courtyard housing exterior, Beijing. Photo by Donia Zhang 1995

The other 14 residents did not comment on the exterior appearance but have addressed other issues related to the form (gate location, sunlight, ventilation, roof design, etc.) via an open-ended question. During the week of handing out and collecting survey questionnaires, I observed four groups of visitors at Juer Hutong taking great interest in the exterior appearance of the housing estate; one said that Juer Hutong was listed in her Guidebook as an attraction site of Beijing.

Housing is always in a process of evolving. Some residents have enclosed their balconies and roof terraces to increase living space, with many windows featuring security grills. A Residents' Committee member recalled that when Professor Wu Liangyong regularly visited Juer Hutong a few years ago, he was disappointed by the balcony enclosures, building extensions, and so on, because these additions altered the original design and made the housing estate look disorderly.

The design of the Tongfangyuan new courtyard-garden housing estate (Case 4) was required to maintain its regional style and features with exterior colors to reflect Suzhou's characteristics of white walls, black-tiled roofs, and brown window and door frames. Detailed planning regulations include a plot ratio of 1:1, exterior form 3-4 storeys, courtyard-style villas with independent doorways and private courtyards, but minimum ornaments and no carvings. Both the project planner and architect indicated that the project has been well received by Suzhou residents who prefer old-style buildings (Qu Weizu, project planner, interview, 2008; Ren Huakun, project architect, interview, 2008).

The planner's and architect's observations confirmed the research findings of Jin *et al.* (2004) that Tongfangyuan residents find the exterior appearance and colors

Figure 7.4 **Juer Hutong new courtyard housing installed with security window grills and balcony enclosures, Beijing. Photo by Donia Zhang 2007**

simple and elegant (44%; n=79), with traces of traditional houses (51%). Residents' perceptions on the aesthetics of Tongfangyuan reached the highest score (14.3; on a scale of +20 to -20), and on local characteristics the second highest (14.07). Residents felt the housing estate holds historic and cultural meaning for them (12.5) and 99 percent appreciated traditional-style housing estate. Jin *et al.*'s random interviews with five residents showed that they all appreciated the designers' efforts to preserve the Jiangnan regional architectural culture (Jin *et al.*, 2004).

The Shilinyuan new courtyard-garden housing estate in Suzhou (Case 5) has been designed in accordance with the city planning regulations: because the site is within a 6-9 m height restricted zone, its style and features must adhere to the regional characteristics and the nearby Suzhou gardens (Peng Hongnian, project architect, interview, 2008). However, only two of 16 interviewed residents commented about the exterior appearance of the housing estate. One resident said that this architectural form complies with the aesthetic value of those attached to traditional Chinese culture. Another said: "I like the exterior appearance of the buildings that looks antiqued and quaint. Many tourists from abroad or other parts of China mistook Shilinyuan as Shizilin [Lion Forest Garden] as they all appreciated its appearance." The other 14 Shilinyuan residents did not comment on the exterior appearance but other issues related to the form (gate location, sunlight, ventilation, roof design, etc.).

Figure 7.5   Tongfangyuan new courtyard-garden housing exterior, Suzhou.
Photo by Qian Yun 2004

The Jiaanbieyuan new courtyard-garden housing estate in Suzhou (Case 6) is located in a 9 m height restricted zone with 3-storey buildings as the main form. Green spaces are between buildings and each town/terraced house has a small, private courtyard. The planning and design have considered the functions of the layouts, adequate landscaping and greening, convenient transportation, and complete service facilities in the surrounding area (Liu Weidong, project planner, interview, 2008).

Nonetheless, only one of 14 interviewed residents at Jiaanbieyuan said that she preferred classical-style housing estate. Although the question was semi-structured, not many residents in Tongfangyuan or Shilinyuan commented about their estate's exterior appearance, either. There is clearly an emerging finding that in each case, apart from a few residents who were enthusiastic about the 'traditional' style of the housing estate, most were indifferent to it. This finding contradicts to that of Jin *et al.* (2004) but conforms to that of Fang (2004, 2006) whose results show that building style is the least important concern when residents purchase an apartment.

Qiu Xiaoxiang, Secretary of Suzhou City Planning Bureau remarked positively that Tongfangyuan, Shilinyuan, and Jiaanbieyuan have strived to combine tradition with modernization by exploring new ways to meet restrictions in building height, form, color, and style, and that they have done well in landscaping and greening (interview, 2008).

Figure 7.6   Shilinyuan new courtyard-garden housing exterior, Suzhou.
Photo by Donia Zhang 2007

Figure 7.7   Jiaanbieyuan new courtyard-garden housing exterior, Suzhou.
Photo by Donia Zhang 2007

Exterior Walls

Although the grey-brick exterior walls of Beijing's renewed/new courtyard housing help preserve its cultural landscape, they are not entirely functional. Four of 13 interviewed residents at renewed traditional courtyard houses (Case 1) report that the depth of the walls had been reduced from 57 cm (or 37 cm) to 24 cm, that the thermal insulation was poorer, and that they perceived it to be more vulnerable to earthquakes. Another two residents complained that nails cannot be drilled in the walls or the bricks will crack. Thus, six of 13 residents noted major problems with the quality of the wall construction.

Figure 7.8   Juer Hutong new courtyard housing exterior walls covered with plants in late summer, Beijing. Photo by Donia Zhang 2007

Seven of 16 interviewed residents at Nanchizi new courtyard housing (Case 2) criticized the exterior walls for their poor thermal insulation, and said the houses were cold in the winter and hot in the summer. Two residents noted that the exterior walls were damp when heating was switched on in the winter as there was a high degree of moisture in them. Moreover, four residents complained that paint peeled off the walls and water seeped through the corners.

The exterior walls of Juer Hutong new courtyard housing (Case 3) are concrete painted white, with visually pleasing plants growing during the warmer months.

However, one of 17 interviewed residents noted cracks on the exterior walls that cause the interiors to be cold in the winter.

All three Suzhou cases have concrete exterior walls in traditional white color. One of 16 interviewed residents at Shilinyuan (Case 5) commented that some exterior walls did not withstand heavy rains, with leaks on the east- and west-facing exterior walls. At Jiaanbieyuan (Case 6), there was no complaint from the 14 residents interviewed about their exterior walls.

Gate and Access

Who can go out but by the door?
How is it that men will not walk according to these ways?
— Confucius, 551-479 BCE, *Analects*, Book 6 Chapter 15

Unlike a traditional courtyard house where a gate served a single-extended family, the ones in housing estates are communal with each estate often having more than one gate for easier access.

Gate Orientation

The survey shows that residents mainly preferred traditional southeast (37%; n=290) for gate orientation, followed by south (14%), and a significant number had no preference (25%). This result suggests that more than half (37%+14% = 51%) of the residents still hold *Feng Shui* principles in gate orientation, with more Suzhou residents (62%; n=123) than Beijing ones (43%; n=167) having these beliefs.

When asked on the survey questionnaire why they preferred a particular orientation for their gate, residents replied that it was traditionally considered lucky to have a gate in the southeast corner because it was the sunrise direction with better light and ventilation, as 'purple *qi* comes from the east,' and that it was nice to walk out into the morning sun (31/290 respondents). Southeast has good *Feng Shui* that may bring prosperity (17/290) and has more luxurious greenery along the river (in Suzhou), close contact with nature (5/290); this direction is also warm in the winter and cool in the summer (3/290). For those who preferred a south orientation for their gate, they contended that since a south-facing gate faces the sun, it is a positive direction (26/290). Some Suzhou residents also mentioned that because their housing estate has a Central Garden, the south/middle gate aided sunbathing and public activities (5/290). Thus, sunlight, wind, and view are the main reasons for residents' preferences for a south-eastern (or southern) gate that is congruent with *Feng Shui* theory. Nonetheless, one respondent chose a north-facing gate because south-facing gate exposes privacy. Two respondents had no preference as long as it was convenient. Due to the small number of respondents who explained their choices, this finding is only indicative.

The Pearson Correlation shows a high, positive correlation between residents' present gate orientation and their preferences. If a resident's present gate faces southeast, his/her preference may be southeast ($r=0.525$; $n=290$; $p<0.000$); likewise, if one's present gate faces southwest, his/her preference may also be southwest ($r=0.596$; $n=25$; $p<0.002$). Since southwest is not a traditional gate orientation, the number of residents who preferred it is low (25). Thus, tradition and habit may influence residents' choices.

Second Security Gate and Screen/Spirit Wall

More residents (54%; $n=290$) considered it *necessary, very necessary, or extremely necessary* to have a second security gate than those who found it unnecessary or extremely unnecessary (46%); there were also more Suzhou residents (55%; $n=123$) than Beijing ones (48%; $n=167$) who thought it was *necessary, very necessary, or extremely necessary* to have a second security gate. This finding implies that the feature of a festooned gate in a traditional courtyard house used as a second security gate is still preferred, but only by some.

The survey shows that more residents (50%; $n=290$) regarded it as *necessary, very necessary, or extremely necessary* to have a screen/spirit wall than those who thought it was unnecessary or extremely unnecessary (44%), and that more Beijing residents (53%; $n=167$) than Suzhou ones (46%; $n=123$) considered it *necessary, very necessary, or extremely necessary*. Nevertheless, differences between the two views are small.

Issues with Gates, Doorways, and Exit

At Nanchizi new courtyard housing (Case 2), each front gate is accessed by the control of a swipe card. A resident criticized the absence of individual door bells and mail boxes, which made it difficult to invite guests or receive mail. One not only had to take time to give visitors directions, but also needed to run out in all weather conditions to open the gate for them. And often, delivered mail got lost.

Since there is no barrier-free access, some households have to make one themselves. Each courtyard compound has an emergency exit, and every household has its own doorway. However, a resident complained that the fire exit next to her house was often used as a regular gate, disturbing her peace.

Juer Hutong new courtyard housing design (Case 3) has incorporated the gradual privacy from semi-public (alleys, paths, and courtyards) to semi-private (stairwells and corridors), to private (apartments) that offers territorial surveillance if trespassers enter the semi-public space (Liu Wenjie, project manager, interview, 2008). However, five of 17 interviewed residents frowned on the new courtyards' numerous passageways that result in much wasted spaces and lost household belongings, with bicycles in particular, when all the doors are unlocked. Although there is a second gate, it cannot be locked because foreign residents tend to come home after midnight, and some with visitors. A resident argued that while the courtyard should be

Figure 7.9   Nanchizi new courtyard housing gates in traditional Chinese
style, Beijing. Photo collage by Donia Zhang 2007

Figure 7.10   The gate to Courtyard A in simplified traditional style at Juer Hutong new courtyard housing, Beijing. Photo by Donia Zhang 2007

accessible in all directions, some doorways were deliberately sealed up by residents and did not function as gateways. This result complies with the survey conducted by Tsinghua University in 1992 that 89 percent (n=31) of its participants found it necessary to lock the courtyard gate at night (Wu, 1999, p. 169).

At Tongfangyuan (Case 4), there are four gates in traditional Suzhou style located in each cardinal direction (east, west, south, and north). Two of five interviewed residents commented positively on the convenience offered by the four gates, and good public safety by the 24-hour security guards.

Shilinyuan (Case 5) has three gates: South, West, and North, and Jiaanbieyuan (Case 6) has two gates: North and South. Each gate has been constructed in traditional Suzhou style and equipped with 24-hour security guards. Several residents at Jiaanbieyuan complained that the long and narrow site of the estate made it inconvenient to walk to public amenities outside the estate because of the distances between individual household doors and the two gates, and that having more gates in the east and west directions would have been helpful. The moon gate on each side of the apartment buildings at both housing estates creates a peaceful green area and connects small and large spaces together. A resident at Shilinyuan affirmed that the space between buildings felt like a *tianjing* ('lightwell'), but it was better than a traditional *tianjing* because the landscape is cared for by property management, making life easier for the residents.

Figure 7.11   A passageway connecting two courtyards at Juer Hutong, Beijing. The residents point out wastage of that space. Photo by Donia Zhang 2007

Figure 7.12   South Gate at Tongfangyuan new courtyard-garden housing estate, Suzhou. Photo by Donia Zhang 2007

Figure 7.13   West Gate at Tongfangyuan new courtyard-garden housing estate, Suzhou. Photo by Donia Zhang 2007

Figure 7.14   South Gate at Shilinyuan new courtyard-garden housing estate, Suzhou. Photo by Donia Zhang 2007

Figure 7.15   West Gate at Shilinyuan new courtyard-garden housing estate, Suzhou. Photo by Donia Zhang 2007

Figure 7.16   North Gate at Jiaanbieyuan new courtyard-garden housing estate, Suzhou. Photo by Donia Zhang 2007

Figure 7.17   A flower gate between apartment buildings in Jiaanbieyuan new courtyard-garden housing estate, Suzhou. Photo by Donia Zhang 2007

Figure 7.18   **A doorway to a town/terraced house with a traditional vertical couplet that reads: "May the good fortune star forever shine on this peaceful home; may prosperity always come to this healthy and happy family"** (福星永照平安宅，好景常临康乐家)**. The horizontal strip reads: "Welcome the bride and receive blessings"** (迎亲接福)**. This household is obviously occupied by newlyweds. Jiaanbieyuan new courtyard-garden housing estate, Suzhou. Photo by Donia Zhang 2007**

Windows

Without opening your door,
you can open your heart to the world.
Without looking out of your window,
you can see the essence of the Dao.

> – Laozi, c.571-471 BCE, *Dao De Jing*, Verse 47

Consider a window: it is just a hole in the wall,
but because of it the whole room is filled with light.

> – Zhuangzi, c.369-286 BCE, adapted by Stephen Mitchell, 2009, Verse 16

The survey shows that residents predominately preferred south (74%; n=290) for their window orientation; the percentage was higher in Beijing (82%; n=167) than in Suzhou (64%; n=123). This answer may be ambiguous because when a respondent chooses 'north,' s/he may actually mean 'sitting north and facing south,' a common *Feng Shui* concept for situating a building. Traditionally, 'north-facing' in China was considered the least favorable because of chilly northerly winter wind from Siberia. This preference for south may be a common practice in the northern hemisphere as a household Chinese phrase goes, "If you have money, do not live in north- or west-facing halls because they are cold in the winter and hot in the summer." Thus, for common-sense interpretation, residents' choices of 'north' and 'south' are combined. This assumption is supported by their explanations in the 'why' section of the survey questionnaire. A significant number of Suzhou residents also preferred an east window orientation (33%).

When asked on the survey questionnaire why they preferred a particular window orientation, residents responded that traditionally, sitting north and facing south gets the best sunlight (132/290 respondents), it is warm in the winter and cool in the summer (22/290), it has positive *qi* and saves energy (2/290), it gives a brighter interior to rouse a better mood (1/290), and it helps the body to absorb calcium (1/290). Those who preferred an east orientation enjoy rising with the sun each morning, a comfort especially in the winter (14/290), and also cultivate better *Feng Shui* because 'purple *qi* comes from the east' and the Chinese phrase goes: "A house in the southeast corner is scarce even with all the wealth to buy one" (11/290). For Suzhou residents, an east orientation is more peaceful, quieter, and more secluded (4/123), with a river view (2/123). A west orientation is also desirable with sunset scenery (2/123). A middle position is good as well because of its proximity to the Central Garden and open space that offers more safety and less noise (4/123). Thus, sunlight, ventilation, and view of nature were the main reasons for residents' orientation preferences, complying with *Feng Shui* theory.

Window Cross-Ventilation Orientations

The study indicates that residents mainly preferred north-south orientation for cross-ventilation (78%; n=290), with this percentage higher in Suzhou (85%; n=123) than in Beijing (74%; n=167). A small number of residents also preferred east-west orientation for cross-ventilation (7%).

When asked on the survey questionnaire why they preferred a particular window cross-ventilation orientation, residents reported that north-south ventilation respected Chinese tradition because China's geographical location determines its prevailing winds coming from the Pacific Ocean in the southeast or Siberia in the northwest, and windows with north-south orientation allow breezes to penetrate rooms more easily. In fact, the bigger the difference between indoor and outdoor temperatures, the better the ventilation, keeping homes warm in the winter and cool in the summer, and bringing comfort in all four seasons (86/290 respondents). Residents share that larger windows facing south are advantageous with more sunlight, and blending light and air brings comfort (31/290). Some respondents stated that these windows create better *Yin Yang* balance and *Feng*

Shui (2/290). A significant finding is that only 2 of 290 respondents explicitly referred to *Yin Yang* and *Feng Shui*, others considered it more of an inherited tradition and unspoken common sense.

The small percentage of residents (7%; n=290) who also selected an east-west orientation for cross-ventilation explained that it is best to have windows in all four (east, west, south, and north) directions to achieve optimal sunlight and cross-ventilation, so that one may open or close any window to facilitate internal airflow (7/290 respondents). The east-west breeze is softer than the north-south wind (4/290), windows in the east and west orientations also allow one to see sunrise and sunset (3/290). Nevertheless, west-facing windows bring in heat in the summer and cold in the winter (1/290). One resident preferred to have skylights in the house to receive direct sunlight.

The study also shows that 52 percent (n=25) of foreign residents in the Beijing sample expressed 'no preference' for their gate orientation, and 56 percent chose 'no preference' for their window cross-ventilation orientation. These data suggest that orientation preference is perhaps a Chinese cultural trait. However, due to the small sample size, this finding is tentative.

Issues with Windows

Five of 13 interviewed residents in Beijing's renewed traditional courtyard houses (Case 1) commented positively that a rear window added to their room brings in more light and ventilation than before. However, two residents complained that after renewal, their rooms have no natural light as there is either no rear window, or it is blocked by the building extensions in the courtyard.

At Nanchizi new courtyard housing (Case 2), seven of 16 interviewed residents observed that their French windows upstairs provide good light and ventilation; one resident also noted that it meets the government regulation of minimum 2 hours of sunlight every day. However, four residents complained that their units are not well ventilated because the houses are built back to back without rear windows. Two residents also said that French windows, uncommon in northern China, are less desirable because they bring in cold in the winter while standard casement windows would have been more functional. In Beijing, the summer afternoon sun is so strong that it can be a problem for people living in west-facing houses, but some residents said their sunshades for west-facing windows were missing. Without verandas, rainwater also comes directly through the frames of doors and windows.

Two of 17 interviewed residents at Juer Hutong new courtyard housing (Case 3) were satisfied with their windows on all four sides (east, west, south, and north) providing plenty of light, all of which could be opened to give full ventilation. However, three residents mentioned that only rooms directly next to the courtyard have good natural air circulation while other rooms do not ventilate well. Some

Figure 7.19   French windows upstairs without a veranda downstairs, Nanchizi new courtyard housing, Beijing. Photo by Donia Zhang 2007

units receive sunlight only after 3 pm, and the windows on the setbacks of the façade have poor sunlight. Residents also complained that 2nd-floor balconies block sunlight to windows on the ground/1st floor. [1] The north-facing balconies are unsuitable for Beijing's climate. The units without balconies are inconvenient for the residents. One resident revealed that although his apartment faces south, it does not get much sunlight; while design regulations require that sunlight should reach 1.5 hours on Winter Solstice (December 21-22), the apartment has clearly not met this criterion. Five residents also complained that the 1st and 2nd floors have poor sunlight, especially for north-facing windows on the 1st floor, due to a close distance between the buildings. Two residents pointed out that the units in the east and west wings are not as enjoyable as those in the south and north wings because west-facing windows bring in hot air in the summer and cold air in the winter, with their observations similar to common speculations in China and *Feng Shui* theory.

Three of five interviewed residents at Tongfangyuan (Case 4) expressed concerns that sunlight and natural ventilation are poor; their living room is dark with the upstairs somewhat better but a lot worse downstairs. This finding confirmed that of Jin *et al.* (2004) who interviewed five residents at Tongfangyuan and they all complained

---

1   British ground floor is same as Chinese 1st floor.

Figure 7.20   The north-facing windows receive poor sunlight, Juer Hutong new courtyard housing, Beijing. Photo by Donia Zhang 2007

about the close building distances, and poor sunlight and natural ventilation. The architectural plans show that Tongfangyuan has an insufficient 10 m distance between the 3-storey buildings. A security guard revealed that poor air quality and foul sewage smells in the area prevents residents from opening their windows.

Five of 16 interviewed residents at Shilinyuan (Case 5) spoke positively that sunlight and natural ventilation are good from the ground/1st to the 3rd floors. Three residents stated that having windows on all four sides brought benefit from sunlight in all directions and internal airflow, with one resident also indicating that having a skylight on the top/3rd floor gains the most direct sunlight. As one resident revealed that since her house has eight windows downstairs and four upstairs, including French windows, it receives excellent sunlight and north-south ventilation. French windows seem suitable to warmer climates of southern China. However, a resident also complained that her ground/1st floor unit gets almost no sunlight for half a year. While some bathroom windows open to the outside and receive natural ventilation, others on the ground/1st and 2nd floors open to public corridors and have poor ventilation, despite fans. While some kitchens are well designed so that cooking exhaust can be emitted directly to the outside, other kitchens without windows have a ventilation problem. While the storage space on the ground floor has good sunlight and ventilation, the underground storeroom is not airy and things go mouldy.

Seven of 14 interviewed residents at Jiaanbieyuan (Case 6) commented that sunlight and natural ventilation are good from the ground/1st to the 4th/top floors, but the north side is gloomy and cold. Some residents specifically choose longer units in the north-south orientation for cross-ventilation. A resident noted that small and staggered windows reduce cross-ventilation. Another observed that ventilation is generally poor for the units in the middle of the buildings where the bathrooms have no windows that open to the outside. Thus, north-facing window orientation, small building distance, and balconies on upper floors all affect the quality of sunlight in a room. Moreover, kitchens and bathrooms should have direct natural ventilation.

Courtyards and Gardens

> *We join spokes together in a wheel,*
> *but it is the center hole that makes the wagon move.*
> *We shape clay into a pot,*
> *but it is the emptiness inside that holds whatever we want.*
>   – Laozi, c.571-471 BCE, *Dao De Jing*, Verse 11

Renewed Courtyards

Residents in Beijing's renewed traditional courtyard houses (Case 1) complained that the renewed courtyards cannot really be called 'courtyards' because there is hardly any yard left. The renewal has not returned them to their original state, but rather, to their more recently filled form. The survey shows that on average 10 households share a renewed courtyard. Six of 13 interviewed residents commented that there are too many households in the compound, and that extensions make the buildings too close to one another. A resident recalled that in the 1950s, she could plant a row of flowers in front of her room, but now there is no space for them. Two residents also bemoaned that it is meaningless to pull down and rebuild a house as it was on the site because the living space has not increased and the courtyard is too crowded. Another resident lamented:

> Due to the small space each household has and the extensions built, the courtyard is changed and distorted beyond recognition. There was originally a screen wall in the courtyard, but it was removed long ago and replaced with a kitchen. The courtyard designs were not flawed; current problems are related to living conditions and not the architecture.

From on-site observations and residents' responses, it is clear that small exterior spaces provided in the renewal projects do not correspond to an understanding or adequate use of a courtyard.

Figure 7.21   A renewed traditional courtyard house at Shejia Hutong with overcrowding, Beijing. Photo by Donia Zhang 2007

New Courtyards and Gardens

The Nanchizi new courtyard housing (Case 2) was designed with 49 new courtyard compounds, each shared by 4-6 households (Lin Nan, project architect, interview, 2008). However, the survey shows that the actual number of households sharing a new courtyard is 4-15, with the average of 7 households sharing a courtyard. The distance between buildings is 7-9 m. Two of 16 interviewed residents were critical that their courtyard is too small for meaningful functions, and that insufficient sunlight makes their home a less desirable place to be. Yet, two other residents felt that the courtyard takes up too much space, and that the boundary between public/private areas is unclear. One also commented on a lack of privacy because her front windows and the neighbors' directly face each other.

Regarding the condition of Nanchizi's new courtyards, three of 16 interviewed residents remarked that it is neat and tidy, that it is good for humans to be in touch with nature, and that it helps to preserve some of Beijing's living traditions. Nevertheless, several residents complained that some new courtyards' landscaping is poor, and there are too many hoarded odds and ends, and that it is visually unattractive. Unruly pets also run around and void themselves everywhere, leaving the courtyard in poor hygiene. The residents suggested a common outdoor storage room, such as a greenhouse for plants and flowers in winter, to help organize the

**Figure 7.22   A new courtyard at Nanchizi, where a vineyard has been built
by the residents, Beijing. Photo by Donia Zhang 2007**

yard space or it will turn into a new 'chaotic-yard.' While inadequate drainage
pipes in the courtyard cause stagnant rainwater, outdoor lighting in the courtyard
for night-time users is absent, despite being essential for the residents who come
home late.

The Juer Hutong new courtyard housing (Case 3) has varied courtyard sizes.
Phase one has a pair of larger courtyards of 13 m × 15 m, each shared by about
15 families in 3-storey apartment buildings, and a pair of side yards (*kua yuan*) of
6.5 m × 7.5 m, each shared by about 4 families in 2-storey apartment buildings.
To raise the floor-area ratio, the courtyard size had to be reduced to result in a far
from ideal form (Wu, 1999, p. 124). Thus, density and plot ratio have negatively
impacted on the courtyard design. Three of 17 interviewed residents criticized the
larger courtyards for being too small, and the (11 m) distance between buildings in
phase two for being too close. As one resident described:

> Our private courtyard is about 4-5 sqm with a tree in it, but it fences off sunlight.
> As the courtyard faces east, sunlight time is very short in the winter, and is only
> suitable for growing plants that do not need much sun, such as ivy, but overall,
> the sunlight is insufficient.

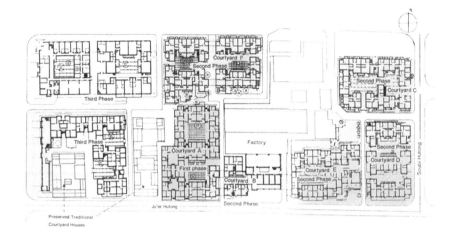

Figure 7.23 Plan of Juer Hutong new courtyard housing showing the three phases and preserved traditional courtyard house, Beijing. Source: Information Center (previously Resources Center) of the School of Architecture at Tsinghua University

Figure 7.24 A new courtyard at Juer Hutong, showing the courtyard used as storage space for cabbage and for drying clothes, Beijing. Photo by Donia Zhang 2007

This finding conforms to that of Liang and Zong (2005). Yet, two other residents at Juer Hutong commented favorably that the new courtyard is good, easy to access, and fresh and green in the summer with many plants. The new courtyard is larger than a traditional one, but also intimately hidden and reserved only for the residents. The children can freely run around and play there, as if on the street, but without the dangers of traffic, cars, and strangers who may harm them.

Due to the constraint of a small site and no design standard for green space at the time (1996), Tongfangyuan (Case 4) does not have much green space between buildings, with a communal garden and kindergarten just 'squeezed-in' (Ren Huakun, project architect, interview, 2008). The site plan reveals that the distances between the 3-4-storey apartment buildings are 10-12 m. Four of five interviewed residents complained that the building distance is too close, a finding that confirms that of Jin *et al.* (2004). In 2008, a homeowner demolished his town/terraced house and built a new one on site with a private courtyard.

The Suzhou Building Code regulates that on the Great Cold Day (January 20-21, the 24th solar division), sunlight penetration should be no less than 2 hours, and on Winter Solstice (December 21-22), sunlight penetration should be at least 1 hour. To meet these requirements, Shilinyuan (Case 5) was designed with 10-11 m distances between the 3-4-storey buildings, with landscaping enclosed by two side walls to create a sense of 'quasi-courtyard.' However, Shilinyuan's site plan reveals uneven building distances that are not always 10-11 m, but they vary between 8-13.5 m. Seven of 16 interviewed residents commented that the 11-13.5 m building distances are wide, and that they like the landscaping. However, two residents observed that the 8-9.5 m building distances are too close.

As Shilinyuan residents are generally aware of land shortage in the old city, they accepted their small courtyards. Two residents living in town/terraced houses revealed that their private courtyards are 30 sqm with the layout perfectly fitting a couple or small family. One resident commented with satisfaction: "My apartment is in a good location, with the communal Central Garden right outside my front door, and a private yard at the back."

The communal Central Garden is a platform for communication because the Suzhou Building Code requires that public green space should be 1 sqm per person, and for groups as large as a 500-resident housing estate, the green space needs to be 0.5 sqm per person, which has determined the size of the Garden. As environmental designs for new housing estates are currently crucial, landscaping is an important aspect at inspection (Peng Hongnian, project architect, interview, 2008).

At Jiaanbieyuan (Case 6), the ratio of building height to distance is 1:1.3, a design based on the Suzhou Planning Regulations for sunlight penetration. The size of the communal Central Garden was determined by the Suzhou City Planning Bureau that it should be no less than 1.5-2 m per person (Liu Weidong, project planner, interview, 2008). Seven of 14 interviewed residents stated approvingly that they enjoyed good sunlight and natural ventilation. Nonetheless, one resident observed that sunlight suffers for some units on the ground/1st floor, adjacent to the 6-7-storey Suzhou High School that blocks the sun.

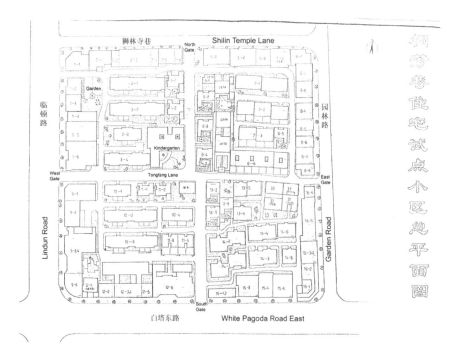

Figure 7.25 Tongfangyuan new courtyard-garden housing estate site plan, Suzhou. Source: Suzhou Urban and Rural Construction Archives, 2007

Figure 7.26 The communal garden at Tongfangyuan new courtyard-garden housing estate, Suzhou. Photo by Donia Zhang 2007

Figure 7.27   The kindergarten at Tongfangyuan new courtyard-garden housing estate, Suzhou. Photo by Donia Zhang 2007

Figure 7.28   Small, private courtyards of town/terraced houses at Tongfangyuan new courtyard-garden housing estate, Suzhou. Photo by Qian Yun 2004

苏州市狮林苑小康小区规划

——12号街坊5号地块总平面图

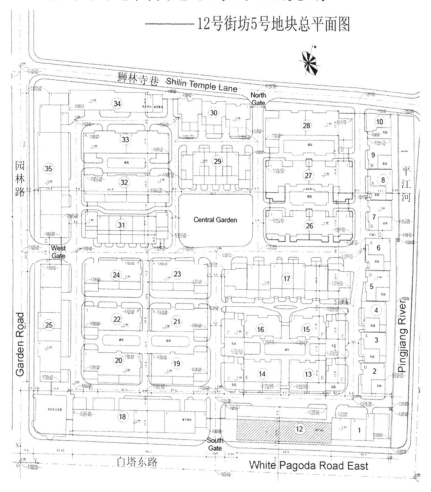

Figure 7.29 Shilinyuan new courtyard-garden housing estate site plan, Suzhou. Source: Suzhou Urban and Rural Construction Archives, 2007

Figure 7.30   The communal Central Garden at Shilinyuan new courtyard-
              garden housing estate, Suzhou. Photo by Donia Zhang 2007

Figure 7.31   A semi-public courtyard at Shilinyuan new courtyard-garden
              housing estate, Suzhou. Photo by Donia Zhang 2007

Figure 7.32   A private courtyard of a town/terraced house at Shilinyuan new courtyard-garden housing estate, Suzhou. Photo by Donia Zhang 2007

Figure 7.33   The communal Central Garden at Jiaanbieyuan new courtyard-garden housing estate, Suzhou. Photo by Donia Zhang 2007

Figure 7.34   A semi-public courtyard at Jiaanbieyuan in Suzhou. Note the
over 6-storey adjacent building that can block sunlight. Photo
by Donia Zhang 2007

Figure 7.35   A private courtyard of a town/terraced house at Jiaanbieyuan
new courtyard-garden housing estate, Suzhou. Photo by Donia
Zhang 2007

Figure 7.36   A well-designed private courtyard-garden of a town/terraced house at Huayanghuayuan ('Sunflower Garden Housing Estate,' 2000, a minor case), Suzhou. Photo by Donia Zhang 2007

Roofs

In the highest antiquity they made their homes (in winter) in caves,
and (in summer) dwelt in the open country.
In subsequent ages the sages substituted for these houses,
with the ridge-beam above and the projecting roof below,
as a provision against wind and rain.

– Yi Jing, Appendix 3 Section 2 Clause 21

The study suggests that residents preferred pitched roofs to flat ones because their experience shows pitched roofs offering better thermal comfort. Although the roofs of Beijing's renewed traditional courtyard houses (Case 1) look like their original form, the reduced tile thickness and eliminated eaves have decreased their protective function. After renewal, four of 13 interviewed residents were critical that their roofs are thinner with poorer thermal insulation. Only after two years of reconstruction, some roof tiles were falling off with leakages when it rains. Moreover, the lack of eaves allows rainwater to slant in.

The Nanchizi new courtyard housing (Case 2) combines pitched roofs with flat ones to express level alternation to enrich the urban scene, when the fixed plot ratio does not allow the design to reduce the floor numbers (Lin Nan, project architect, interview, 2008). It should be pointed out that although some buildings in classical Chinese courtyard houses had flat roofs, these buildings were used only as service rooms; all the buildings for human habitat had pitched roofs (Ma, 1999).

Seven of 16 interviewed residents at Nanchizi indicated that flat roofs are not as functional as pitched ones because their houses are hot in the summer and cold in the winter, with a temperature difference as much as 3°- 4°c between the ground/1st and 2nd/top floors. One resident noted that visually, flat and pitched roofs built side by side are unsymmetrical and unaesthetic. Moreover, the roof tiles have been insecurely glued on with mortar continually falling off, losing the roof's protective status. Furthermore, the eaves are shallow with rainwater slanting in, as traditional eaves were typically 60 cm for blocking rainwater.

At Juer Hutong new courtyard housing (Case 3), one of 17 interviewed residents revealed that her roof terrace leaked in the rain for over 10 years, but it cannot be fixed because it is a design flaw. Another resident said that her roof is only pitched on one side, with a big, deep hollow like a swimming pool, although aesthetic but impractical, as a few tree leaves can block the drainage when it rains. Moreover, all the rooms on the roof level are hot in the summer.

As traditional Suzhou houses have pitched roofs at a slope between 4:12 and 6:12 (Suzhou Housing Management Bureau, 2004), Tongfangyuan (Case 4) was built with pitched roofs at a slope of 4:12. No resident at Tongfangyuan, Shilinyuan (Case 5), or Jiaanbieyuan (Case 6) complained about their roofs.

Overall, only three of 82 interviewed residents mentioned that their roof terraces are desirable, and eight residents commented that pitched roofs offer better thermal performance. Thus, residents have responded more in terms of thermal comfort than a cultural perception or aesthetic value.

Summary and Conclusion

This chapter discussed the form and environmental quality of the renewed/ new courtyard housing in Beijing and Suzhou. The findings show that although Beijing's renewed traditional courtyard houses (Case 1) have maintained the appearance of the old city, the unsystematic planning has weakened its effect where integrated planning may have had better consistency of exterior forms. Unique and unconventional, the Nanchizi new courtyard housing (Case 2) resembles neither Beijing's traditional courtyard houses nor Western-style villas. The Juer Hutong new courtyard housing (Case 3) has imitated southern Chinese vernacular architecture but has not retained local or regional identity based on on-site observations. On the other hand, Tongfangyuan (Case 4), Shilinyuan (Case 5), and Jiaanbieyuan (Case 6) have adapted Suzhou's vernacular architectural style and features with relative success, and have preserved a sense of local/regional identity.

This study finds that while some residents value the traditional-style exterior appearance, most were indifferent to it. While the exterior walls in traditional

Figure 7.37 A flat roof at Nanchizi new courtyard housing: the lack of eaves allows rainwater to slant in and the rooms are hot in the summer. Photo by Donia Zhang 2007

Figure 7.38 A roof pitched only on one side, Juer Hutong new courtyard housing, Beijing. Source: Information Center (previously Resources Center) of the School of Architecture at Tsinghua University

materials and colors were appreciated, the new construction methods employed in some cases have meant that exterior walls do not always function well.

The study also shows a significant number of residents still valued traditional southeast (or south) gate orientation and north-south cross-ventilation orientation, which confirms some aspects of *Feng Shui* theory. Some residents preferred having a window in all four directions to gain maximum sunlight, ventilation, and views of sunrise and sunset. If there is only one window, most residents still preferred it to be south-facing (only 6 percent of Chinese respondents had no preference). While these preferences relate to a perceived better environmental quality, this perception may be culturally determined.

The renewed/new courtyards are generally too small to admit enough sunlight. The architectural drawings show that the new courtyard proportions are no longer preserved as in tradition. To achieve the same amount of sunlight as in traditional courtyard houses, the ratio of building height to distance should be at least 1:3 for Beijing (Zhang, 1994, 2006, 2009/2010/2011) and 1:1.3 for Suzhou, which means a minimum of 18 m distance for 6 m high surrounding buildings in Beijing, and 12 m distance for 9 m high surrounding buildings in Suzhou. However, the three Beijing cases and two Suzhou cases have not met these criteria, which have seriously affected their environmental quality.

Residents have expressed a preference for pitched roofs rather than flat ones because they have experienced better thermal performance of pitched roofs.

The findings indicate that the physical attributes of the new courtyard housing are not always in harmony with heaven, and that they cannot provide residents maximum physical protection or comfort under various weather conditions, even though the Suzhou cases generally fare better than the Beijing ones. As well, the symmetrical planning along a central axis linking Heaven and Earth in traditional Chinese philosophy has not been observed in the design of Nanchizi (Case 2), Juer Hutong (Case 3), or Jiaanbieyuan (Case 6).

In Chapter 8 that follows, the space and construction quality of the renewed/ new courtyard housing will be explored, to see whether they are in harmony with earth, and whether they satisfy residents' living requirements.

Table 7.1   Summary of the form of the renewed/new courtyard housing

Design Elements	Beijing Cases			Suzhou Cases		
	1. Renewed traditional courtyard houses	2. Nanchizi new courtyard housing	3. Juer Hutong new courtyard housing	4. Tongfangyuan new courtyard housing	5. Shilinyuan new courtyard housing	6. Jiaanbieyuan new courtyard housing
Exterior form	Traditional	Somewhat traditional	Southern traditional	Somewhat traditional	Somewhat traditional	Somewhat traditional
Exterior walls	Grey bricks	Grey bricks	White concrete	White concrete	White concrete	White concrete
Gate orientation	Southeast	Varied	Southeast or south	South, north, east, and west	South, north, and west	North and south
Window orientation	Traditional	Untraditional	Untraditional	Untraditional	Untraditional	Untraditional
Courtyards	Communal	Communal and private	Communal	Communal and private	Communal and private	Communal and private
Roofs	Grey-tile pitched	Grey-tile pitched mixed with flat ones	Grey-tile pitched, some with terraces	Black-tile pitched	Black-tile pitched, some with terraces	Black-tile pitched, some with terraces

Source: My summary

Chapter 8
Harmony with Earth:
Space and Construction Quality of the
New Courtyard Housing

In dwelling, live close to the ground.
In family life, be completely present.

– Laozi, c.571-471 BCE, *Dao De Jing*, Verse 8

We hammer wood for a house,
but it is the inner space
that makes it liveable.
We work with being,
but non-being is what we use.

– Laozi, c.571-471 BCE, *Dao De Jing*, Verse 11

This chapter analyzes the findings on the quality of space and construction in the renewed/new courtyard housing in Beijing and Suzhou. It investigates whether the spatial designs satisfy residents' living requirements, and whether the building materials and construction methods are in harmony with earth, which is a fundamental of Chinese philosophy (Chapter 3), and a cornerstone of culturally sustainable architecture in China (Chapter 11). The spatial and constructional aspects studied include interior space, floor levels, furniture styles and materials, facility provision, building materials and construction quality, maintenance and management, and car park spaces.

Interior Space

Beijing's renewed traditional courtyard houses (Case 1) have not solved the problem of small living space. Although three of 13 interviewed residents commented positively that their house condition is much better than before, the rooms are brighter, the house tidier, and the environment cleaner, five residents still preferred larger rooms because demolition and reconstruction on site basically maintained the size of original living spaces. Some houses are only slightly enlarged (4 sqm) by reducing the thickness of exterior walls, resulting in poorer sound and thermal insulation, and more danger in case of an earthquake. The ceiling is also lowered, making residents feel compressed and the airflow worse than before.

Moreover, the renewed traditional courtyard houses still lack private bathrooms. Although some courtyards had a lavatory before the 1960s, this was replaced by public toilets in the *hutong*. Self-built kitchens and storage spaces in the courtyard still exist after renewal, reducing the courtyard to a narrow passageway. Despite the good intentions of Beijing municipal government, renewal has not actually improved the residents' living conditions.

Two of 13 interviewed residents suggested building semi-basement kitchens and bathrooms to create more space for the courtyard. Seven residents recommended 2-3-storey new courtyard housing on site, with a private kitchen and bathroom in each house, one semi-basement and two levels above ground, and a yard shared by 4-5 households. Two residents expressed a willingness to spend their own money to reconstruct their own courtyard house on site, which conforms to Zhao's finding (2000, p. 96).

Three of 16 interviewed residents at Nanchizi new courtyard housing (Case 2) commented that their living condition has been improved compared with that in dilapidated traditional courtyard houses on site because now there was a kitchen and a bathroom in each housing unit. Although some residents initially felt uncomfortable about the 2-storey interior space, they have adapted to it.

However, six residents at Nanchizi new courtyard housing observed that the layout lacks rationality and each unit has a different design. Twenty of 71 (28%) survey respondents criticized the rooms in a 2-storey house of 45-60 sqm for being too small, and the kitchen and bathroom with a staircase leaving only a passageway but hardly any hall on the ground/1st floor, with almost no space for a dining table. In some houses, the living room's width makes it difficult to accommodate a TV, and many houses have no living room at all. The staircase is also too narrow, steep, unsafe, and uneasy for moving furniture (12/16 respondents). The kitchen and bathroom are too small (7/16) as the bathroom is only 1-2 sqm with no room to turn around for taking a shower or installing a washing machine. The lack of a bathroom upstairs creates inconvenience at night, especially for the elderly. The staggered layouts are dysfunctional as it is spatially inefficient and uncomfortable (6/16). Lastly, there is no balcony for drying laundry (2/16). A resident described this inconvenience:

> The space inside is generally cramped. In the beginning, I found it kind of quaint, but as years went by, I got tired of bumping into things all the time. I would never buy a house such as one in Nanchizi. The only value these houses have is their location. The entire interior is an architectural failure.

Another Nanchizi resident suggests that to alleviate the problem of tight living space, a basement should have been built. However, two other residents contend that a 2-storey house is inconvenient for the elderly and physically impaired, who should live on the ground/1st floor.

Figure 8.1   The interior space of a traditional courtyard house has not increased after renewal, Qingfeng Hutong, Beijing. Photo by Donia Zhang 2007

Figure 8.2   Some renewed courtyards only have dangerously narrow passageways because of self-built extensions, Juer Hutong, Beijing. Photo by Donia Zhang 2007

Figure 8.3   The downstairs of a 1-bedroom house with Western-style interior space, Nanchizi new courtyard housing, Beijing. Photo by Donia Zhang 2007

Figure 8.4   The upstairs of a 1-bedroom house, Nanchizi new courtyard housing, Beijing, where some residents combine the bedroom with an office. Photo by Donia Zhang 2007

Figure 8.5   A kitchen with modern stove in a 1-bedroom house, Nanchizi new courtyard housing, Beijing. Photo by Donia Zhang 2007

Three of 17 interviewed residents at Juer Hutong new courtyard housing (Case 3) observed that some units' spatial design is irrational since each unit has a different layout except those in the north and south corners. Rooms facing the small side yards (*kua yuan*) are shaded all year round, humid, and uncomfortable. Five residents noted that the units of 40-60 sqm are too small with 5-6 doors leading to the hall, but no living room in some units, making it difficult to arrange furniture. The bedrooms are tiny, since one has to sit on the bed as soon as entering the room. The 2-sqm kitchen is too small for a fridge and the 2-sqm bathroom is too cramped for taking a shower or installing a washing machine (4/17 respondents). The 2.5m-high ceiling creates a feeling of constraint while 2.8m- to 3m-high would have been better (4/17). The floor is too thin as even a tiny pin drop on the 2nd floor can be heard on the ground/1st floor. Such noise affects residents' health, especially those with heart disease. For duplex apartments on the 3rd and 4th floors, the bedrooms are upstairs without a bathroom, which is inconvenient at night. Some duplex apartments have staircases located right in the middle of the hallway, causing both a waste of space and awkwardness for arranging furniture (5/82). The steep stairs can be dangerous for both children and the elderly (4/17). Although two residents have expanded their units by enclosing the private yard, balcony, or roof terrace, these additions create visual disorder.

Nevertheless, the Juer Hutong new courtyard housing did fulfil residents' basic living requirements at the time, although it may no longer suffice. Current living requirement in the inner city is that each unit must be 100-180 sqm, which far exceeds the standards back then (Liu Wenjie, project manager, interview, 2008; Wu Chen, son of Professor Wu Liangyong, interview, 2008).

Some apartments at Tongfangyuan (Case 4) are smaller than the 90 sqm requirement set by the Suzhou City Planning Bureau, and two of five interviewed residents complained about their small unit sizes of 70 sqm and 75 sqm. This finding echoes that of Jin *et al.* (2004) that the major spatial design problems of Tongfangyuan are irrational layout, small units, and low ceilings.

Since Shilinyuan (Case 5) was designed at a time when the fashion was to build large living rooms but small bedrooms, several residents indicated that some units have bedrooms big enough only for a bed and two bedside tables, with no room for a chair; one has to awkwardly sit on the bed after entering the room. Three of 30 interviewed residents at Shilinyuan and Jiaanbieyuan (Case 6) preferred more spatial divisions for practicality, especially more bedrooms if an elderly or a sick person needs a living-in nanny. Moreover, an apartment with several staggered levels connected by steps makes it inconvenient to go between the floors. In Suzhou, the ground-/1st-floor apartments are generally damp during plum rains in June and July (in the middle and lower reaches of the Yangzi River), but these new houses are not raised above ground to block rainwater. While traditionally Suzhou has no basements because of a moist and humid climate, a resident observed that underground storage is still necessary for bicycles and household objects.

Although the research did not set out to find the optimum unit size, some residents in Beijing Nanchizi (Case 2) and Juer Hutong (Case 3) mentioned that a unit of 60 sqm is barely enough for a couple, and it feels crowded with visitors (2/33 respondents). A unit of 70-90 sqm is comfortable as there is no problem hosting friend (2/33). A unit of 100-120 sqm is large and reasonably satisfactory (2/17). Suzhou Tongfangyuan (Case 4) residents commented that while a unit of 70-75 sqm is too small and awkward, 80 sqm is acceptable for 2-3 persons (3/5). While Shilinyuan (Case 5) and Jiaanbieyuan (Case 6) have larger and similar unit sizes, residents observed that a unit of 90-120 sqm is sufficient for a 2-3-person household (7/30) and 120-150 sqm is very satisfactory (7/30), they also found 150-180 sqm to be very spacious (5/30) and 220-350 sqm very large (3/30). Thus, a unit of 90-180 sqm is generally a satisfactory interior space. Due to the small number of participants mentioning their unit sizes, these findings are only suggestive.

Similarly, Fang's (2004, 2006) research results show that when purchasing an apartment, Beijing residents' priorities are unit size, efficient unit layout, and spacious public areas.

Figure 8.6   Although the roof-level enclosure adds interior space, it creates visual disorder, Juer Hutong new courtyard housing, Beijing. Photo by **Donia Zhang 2007**

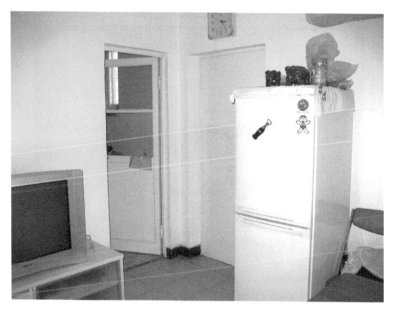

Figure 8.7   One side of a hall with two doors leading to it, Juer Hutong new courtyard housing, Beijing. Photo by **Donia Zhang 2007**

Figure 8.8   **A sitting room on the ground/1st floor of a town/terraced house with traditional ornaments, Shilinyuan, Suzhou. Photo by Donia Zhang 2007**

Figure 8.9   **A spacious bedroom on the 2nd floor of a town/terraced house, Shilinyuan new courtyard-garden housing, Suzhou. Photo by Donia Zhang 2007**

Figure 8.10 Traditional Chinese-style interior decoration in a Western-style spatial layout of a town/terraced house, Jiaanbieyuan new courtyard-garden housing, Suzhou. Photo by Donia Zhang 2007

Floor Levels

The research reveals that residents preferred 1-3-storey housing: ground/1st floor (41%; n=290), 2nd floor (37%), and 3rd floor (32%). After that level, their preferences dramatically dropped (8%).

A difference in floor level preference is noted between Beijing and Suzhou respondents, arguably due to climatic factors. Beijing residents mainly preferred the 1st (53%; n=167) and 2nd (24%) floors, and had less preference for the 3rd floor (16%). Beyond that, their preference radically declined (3%), with only a small number of them who preferred a basement (2%).

When asked on the survey questionnaire why they preferred a particular floor level, Beijing respondents reported that the ground/1st floor has courtyards for parking bikes and storing household goods (24/167 respondents), or for planting flowers and drying laundry in the sun (17/167). Earth *qi* and sunlight seem important to their health (12/167), and that having activities on the ground/1st floor in the summer and 2nd floor in the winter is convenient, especially for the elderly (7/167). If single-storey houses can solve the problems of a lack of heating and bathroom, they are seen as more advantageous than multi-storey buildings (3/167). Residents who preferred the 3rd floor stated that it has duplex apartments (3/167) with better sunlight (5/167) and ventilation (3/167), it is neither too high nor too low (3/167), it offers a better view (3/167), it maintains a cleaner environment (3/167), it gives more privacy (2/167), and it features a roof terrace (2/167). Thus, outdoor space, sunlight, natural ventilation, and earth *qi* seem to be the main considerations when residents choose a floor height. However, due to the small number of respondents who explained their choices, this finding is only indicative.

A resident at Nanchizi (Case 2) new courtyard housing said that he likes sleeping downstairs more than upstairs, perhaps because of having previously lived in a traditional courtyard house for 25 years. Several Nanchizi residents indicated that living on the ground/1st floor is better for the elderly, reflecting a desire for traditional living. However, several residents at Beijing's renewed traditional courtyard houses (Case 1) contradicted the finding by suggesting rebuilding 2-3-storey new courtyard housing on site.

Suzhou respondents mainly preferred the 3rd (55%; n=123) and 2nd (37%) floors more than the 1st (24%). Above the 3rd floor, their preference significantly decreased (15%).

When asked on the survey questionnaire why they preferred a particular floor level, Suzhou residents replied that the ground/1st floor has courtyards for gardening and a view (21/123 respondents), and it is convenient, especially for the elderly (18/123). Connected to the earth *qi*, the ground/1st floor is similar to living in traditional courtyard houses (4/123) although this floor may be damp, dark, and noisy (13/123), with mosquitoes (1/123). Others responded that the ground/1st floor can be used for living and 2nd floor for sleeping (10/123). Those who preferred the 2nd floor argued that it is less damp (7/123) than the 1st floor, and less hot than the 3rd/top floor in the summer (5/123). Those who preferred the 3rd/top floor reported that it has better sunlight and natural ventilation (24/123), better views (3/123), drier air (4/123), it is quieter and not too high (9/123), and convenient even if there is no elevator (5/123). These findings suggest that sunlight, ventilation, and view are important considerations when selecting a floor level.

The Pearson Correlation indicates that there is a low, positive correlation between the residents' present floor level and their preferences. If they currently live on the ground/1st floor, they may prefer to live on the ground/1st floor (r=0.325; n=285; p<0.000); and if they presently live on the 2nd floor, they may also prefer to live on the 2nd floor (r=0.254; n=285; p<0.000). Therefore, habit seems to affect their floor preferences.

My findings conform to a MPhil study by Zhao (2000) who surveyed 112 households and interviewed 20 residents living in four housing types: one-storey traditional courtyard houses, low-rise housing, mid-rise, and high-rise apartment buildings in Beijing's Eastern District (*Dongsi*). His survey results indicate that residents preferred traditional courtyard houses to the other three housing types, and that courtyard space is rated as high as 72 percent as a major aspect of their satisfaction. Among those living in courtyard houses, 53 percent favored them, and among those who had previously lived in them but now live in apartment buildings, 48 percent still preferred the courtyard house. Zhao's findings also suggest that between high- and mid-rise apartment buildings, mid-rise is rated higher (34%) than high-rise ones (22%). On average, the courtyard house plan is rated the highest (43%) of the three housing types, followed by the 'slab' plan (22%) and the U-shaped plan (21%); the free-standing tower plan widely used in modern residences received the lowest rating (10%) (pp. 94-95). Many residents also indicated that 'gaining *qi* from the earth' (*Feng Shui*) is a valuable asset found in traditional courtyard houses (pp. 112-113).

My results also support Wu Liangyong's (1999) argument that, for Beijing's historic residential areas, low building heights featuring mainly courtyard housing of 2-3 storeys should be designed with sub-basements (p. 201) because people prefer this floor height. If the buildings surrounding a courtyard are above 3 storeys, light into the courtyard becomes limited, making the space less desirable.

Furniture Styles and Materials

Four distinct styles of furniture are observed across the six estates: classical Chinese style (27/67 respondents), modern Western style (18/67), modern Chinese style (9/67), and mixed Chinese-Western style (9/67). An additional number of households have no discernible style of furniture (4/67). Four of 67 interviewed residents showed an environmental awareness when choosing materials for their furniture. For example, one said, "We have solid wood furniture in a simple, modern style mainly for the consideration of environmental protection and simplicity." Another resident revealed, "My home decoration is simple, cheerful, and the material is environmentally friendly."

Residents who chose classical Chinese-style furniture made of red/hard wood explained that their choice matches the style of their house or housing estate (17/67 respondents), that it looks and feels rich, tasteful, and durable, and it has a store of value (8/17), making their choice more connected to traditional Chinese philosophy. Residents who selected modern Western-style furniture did so because of the lack of living space (4/67) or because they were fond of a simple and comfortable lifestyle (3/67). As modern Western-style furniture is often made of varied synthetic materials such as plywood, panel board, composite board, high-density board, or plastic veneer board, it is valued for its cost (8/67) and attractive style (5/67), and because it is easy to clean (3/67), simple to transport (2/67), and uncomplicated to assemble (2/67).

A resident explained that because her family manufactures and sells Chinese-style furniture, it would be boring to also display this style at home. Another resident described her furniture as solid wood in modern style chequered with black and white (*yin* and *yang*) contrasting colors resembling classical Suzhou architecture. Still another Nanchizi resident explained, "My furniture is a blend of Chinese with Western styles because the new courtyard housing features some kind of Western form while showing Chinese paintings on the beams and columns in the courtyard."

A Juer Hutong resident observed that the preference for furniture style has a connection with age. Elderly people tend to prefer solid-wood furniture in classical Chinese style, while younger generations often choose plastic veneer-board furniture more for appearance than durability. Thus, classical Chinese-style furniture was the most common of the four styles (40%; n=67), a finding that contradicts that of Kai-Yin Lo (2005) in *Traditional Chinese Architecture and Furniture*: "This is a clear indication of the rejection of the old, as well as a lack of appreciation of cultural value and a slavish adoption of the new. In more affluent urban households, the trend is to follow Western décor and fashion" (p. 203). While some residents in the new courtyard housing may be affluent, this study shows no obvious *trend* of preferring Western-style furniture.

Facility Provision

> *He who aims to be a man of complete virtue in his ... dwelling place*
> *does not seek the appliances of ease; he is earnest in what he is doing.*
> — Confucius, 551-479 BCE, *Analects*, Book 1 Chapter 14

Traditional Chinese courtyard houses were neither equipped with 'comfortable' facilities by Western standards (Knapp, 2005b), nor were they intended to be so, because Confucian ethics emphasize frugality towards material life, and now modernity demands a change to it. But if modernization aims at the betterment of human existence, there is no fundamental contradiction between major Confucian principles and modernization (Lau, 1991, p. 222).

Three of 13 interviewed residents at Beijing's renewed traditional courtyard houses (Case 1) commented that certain facilities were improved. Every household had an electricity and water meter now, to avoid neighbors' quarrels over utility bills and to reduce water usage because residents are more careful with using water when they have to pay for it themselves. A resident further suggested flushing the toilet with grey or used water to save for the future. However, the renewal has not provided a modern heating system: the houses still relied on coal burning at the time of the survey (2007), exposing residents to the risk of carbon monoxide poisoning (4/13 respondents). Several households have self-installed heating and air-conditioning, but if they switch on air-conditioner in the summer, the brake may switch on due to the insufficient power capacity (2/13). Thus, the renewal has not solved the problems of facility inadequacies.

Three of 16 interviewed residents at Nanchizi new courtyard housing (Case 2) indicated that although the new housing fulfills their needs for basic living facilities, the heating system uses wall-mounted units that not only waste energy in the winter, but also are not warm and are more costly.

The basic facilities at Juer Hutong new courtyard housing (Case 3) are either crude or missing. Unaesthetic and unsafe pipes are exposed in each unit. Although hot water comes from underground, four of 17 interviewed residents complained that their heating system is neither functional nor reparable due to a backwater design flaw. The rooms can get frigid in the winter, especially the north-facing

Figure 8.11   The road of Juer Hutong after reconstruction is higher than the courtyard level, forcing rainwater to flow back into the yards. Photo by Donia Zhang 2007

ones. Moreover, there is no space for installing an energy-efficient solar heater (2/17 respondents). Due to their narrow diameters, the ground/1st floor kitchen and bathroom pipes often get blocked, with backed-up sewage and annoying sounds of water running in the pipes (3/17). With no gas pipes installed in these buildings, residents have to refill gas tanks on a regular basis. At the time of the survey (2007), the road of Juer Hutong was being reconstructed, but gas pipes were still uninstalled. Whenever the city installs new service pipes, the *hutong* grade level is raised higher than the courtyard level, forcing rainwater to flow back into the courtyards, which is slippery for residents to walk on (Residents' Committee members, interview, 2007).

Planning regulations for Tongfangyuan (Case 4), Shilinyuan (Case 5), and Jiaanbieyuan (Case 6) mandate that basic services such as water, electricity, and gas, be provided (Liu Weidong, project planner, interview, 2008; Qu Weizu, two projects planner, interview, 2008). While residents at the three Suzhou cases seemed to be satisfied with their facility provisions, one of five interviewed residents at Tongfangyuan complained about stagnant sewage water and deposited garbage from surrounding restaurants and hotels that compromise the housing estate's hygiene.

One of 16 interviewed residents at Shilinyuan objected to disturbing noise generated from her neighbor's air-conditioner. Two residents were critical of the heat emitted from their neighbor's air-conditioners because it discouraged them from opening windows. Another two residents reported that when their small sewer pipes are blocked with backed up drainage water, their kitchen and toilet will also be flooded. Only one of 14 interviewed residents at Jiaanbieyuan commented approvingly on the advanced kitchen design, as a chimney emits lampblack out to the roof.

Building Materials and Construction Quality

> *If a man takes no thought about what is distant, he will find sorrow near at hand.*
> – Confucius, 551-479 BCE, *Analects*, Book 15 Chapter 11

The poor building materials and substandard construction quality in Beijing's renewed/new courtyard housing have caused more anxiety to residents than their enjoyment and use of the courtyards.

To enforce construction regulations, China's Ministry of Housing has issued *Residential Design and Construction Specifications* (1987, 2003, 2006), *Civil Design General* (1987, 2005), *Architectural Design Code for Fire Protection* (1987, 2006), among others, to enact them in every city, along with a Construction Documents Inspection Center to uphold these guidelines. Andrew Smeall has observed that China has fairly rigorous building codes, but what is in the policies and what is followed and practiced are sometimes two different stories (Smith, 2008).

Materials and Construction of Renewed/New Courtyard Housing in Beijing

Three of 13 interviewed residents at Beijing's renewed traditional courtyard houses (Case 1) commented positively that the quality of reconstruction is excellent and that the environment is improved. However, seven residents were critical that the present renewal is hasty, careless, and seriously substandard without long-term plans. Four residents observed that their walls crack easily because of poor-quality cement, or no cement at all between bricks. Rain also leaks into renewed but thinner gypsum ceiling boards (2/13 respondents). Moreover, ceiling paint peels off and putty on door and window frames also falls off (3/13). The floor is unlevelled with poorly layered cement that crusts into dust and debris underfoot (2/13). The rush for completion at the expense of quality of reconstruction results in a nuisance for post-renewal repairs.

Regarding the materials for Nanchizi new courtyard housing (Case 2), the lead developer from Eastern District Housing and Land Management Center contended, "Under the circumstances that traditional materials can still comply with modern functional requirements, it is best to use traditional building materials and construction techniques." Nevertheless, the lead developer did not answer my further questions concerning the quality of material and competence of workmanship.

Figure 8.12 Cement that crusts into dust and debris underfoot, renewed traditional courtyard house, Qingfeng Hutong, Beijing. Photo by Donia Zhang 2007

Figure 8.13 A column is cracked from inside, renewed traditional courtyard house, Qingfeng Hutong, Beijing. Photo by Donia Zhang 2007

Figure 8.14 **A column is cracked on the outside, renewed traditional courtyard house, Qingfeng Hutong, Beijing. Photo by Donia Zhang 2007**

Figure 8.15 **Unsupported by the doorframe, a beam's weight rests on the window glass, renewed traditional courtyard house, Qingfeng Hutong, Beijing. Photo by Donia Zhang 2007**

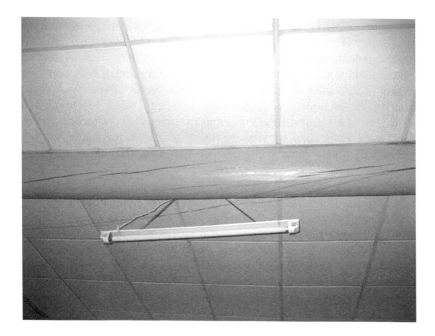

Figure 8.16   A beam is cracked from inside, renewed traditional courtyard
house, Qingfeng Hutong, Beijing. Photo by Donia Zhang 2007

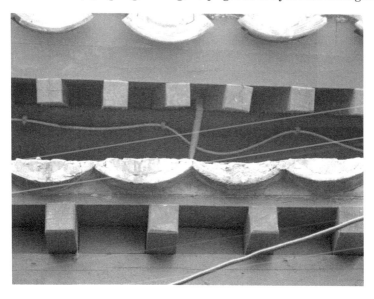

Figure 8.17   Roof tiles are filled in with cement, renewed traditional
courtyard house, Qingfeng Hutong, Beijing. Photo by Donia
Zhang 2007

Seven of 16 interviewed residents at Nanchizi new courtyard housing noted that the quality of construction is generally poor; it was hastily completed without much attention to details in workmanship. The iron doors and windows are not airtight and rain leaks into the house (2/16 respondents), with poor sound insulation in a location so close to the Nanchizi Street (3/16). Two residents showed a preference for doors and windows made of wood. Moreover, as the window glass is made of deficient materials to reduce the construction cost, it is thin and traps water vapor between the double-glazing in the winter, fogging it up and making it impossible to clean (2/16). Two residents request to have their houses rebuilt. A resident writes:

> Nanchizi was admittedly an architectural failure that local government officials have apparently acknowledged. That is why Beichizi has not been rebuilt the same way (this had been the original plan). If you actually live in one of the houses, you will realize what poor quality building, poor insulation and ventilation can be in practice. The mobile phone signals are dreadful inside many housing units [of course, this is not the builder's fault], forcing residents to use their phones out in the yard that disturbs others.

Figure 8.18   Self-installed thermal insulation boards by residents, Nanchizi new courtyard housing, Beijing. Photo by Donia Zhang 2007

Figure 8.19   Water vapor unable to escape from a double-glazed window, making it impossible to clean, Nanchizi new courtyard housing, Beijing. Photo by Donia Zhang 2007

Figure 8.20   **Damp smudges and wall blisters beside a window frame, Nanchizi new courtyard housing, Beijing. Photo by Donia Zhang 2007**

Figure 8.21   **A window's rotted wooden-frame only after 4 years of construction, Nanchizi new courtyard housing, Beijing. Photo by Donia Zhang 2007**

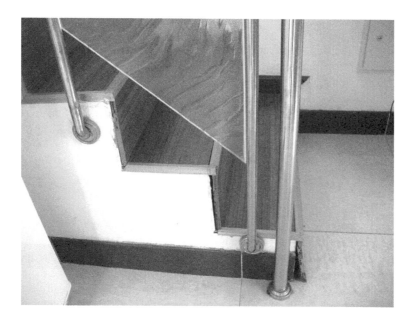

Figure 8.22 A crudely constructed staircase, Nanchizi new courtyard
housing, Beijing. Photo by Donia Zhang 2007

Figure 8.23 A resident-invented method of cold prevention around the
window, Nanchizi new courtyard housing, Beijing. Photo by
Donia Zhang 2007

Thus, Nanchizi residents may be less able to appreciate some positive attributes of the new courtyard housing because of its substandard quality of materials and construction.

At Juer Hutong new courtyard housing (Case 3), some units have poor soundproofing. A resident complained that they can hear almost everything the next-door neighbor is up to. While weak sound barriers compromise music practices, mobile/cell phone signals can be undetectable. Liang and Zong (2005) also found that Juer Hutong's substandard construction quality has rendered it to rapid decline.

Political Agenda and Construction Quality

Lin Nan, the project architect of Nanchizi new courtyard housing (Case 2) revealed that, to finish the project before National Day (October 1), which is a common practice in China, the construction document was completed in just one month. The construction work was carried out at the same time as drawings were drafted, with a new set of drawings delivered to the site every week. The whole project was finished in less than 3 months. If there is an issue with the construction quality, it was a direct result of the time limit (interview, 2008). Moreover, it appears that construction quality assurance was also missing for the project.

Materials and Construction of New Courtyard-Garden Housing in Suzhou

Shilinyuan new courtyard-garden housing (Case 5) has applied new materials to save energy and resources. For example, load-bearing perforated clay bricks (30% hole) that offer good thermal insulation were used instead of solid clay ones. With well-insulated doors and windows, the design can save 50 percent of energy consumption (Peng Hongnian, project architect, interview, 2008). A resident at Shilinyuan observed that the walls are firm and sturdy, with adequate sound and thermal insulation. However, some east- and west-facing exterior walls leak during thunderstorms.

One of five Tongfangyuan (Case 4) residents reported that the sound insulation is poor, but there were no complaints from Jiaanbieyuan (Case 6) residents about its construction quality.

Maintenance and Management

The institutional management procedures that residents have to deal with at the renewed/new courtyard housing would traditionally have been decided either within a family or through an agreed code of ethics.

Public Housing Management

A government-owned organization, the Housing Management Office, is responsible for resolving issues arising from Beijing's renewed traditional courtyard houses (Case 1), and three of 13 interviewed residents reported with satisfaction that the reconstruction team repaired their leaking roofs without extra charge.

However, the public management office is not performing well at Nanchizi new courtyard housing (Case 2). A resident revealed that it is not really property management, but run by the previously government-owned Housing Management Bureau with untrained and unqualified staff. No formulation, regulation, or measurement is in place for maintenance fees, although maintenance work is carried out once every three years. In 2007, all the exterior walls and decorations in the courtyards were repainted using maintenance fees collected from residents. Nevertheless, maintenance work is not consistently conducted. Three of 16 interviewed residents complained that the paint on the façade has faded and peeled off, but nobody tries to retouch it. When roof tiles fell off a few years ago, the property management asked the construction team to repair them. Now the team has left, nobody restores them. The gate knocker is damaged and dead tree branches fall in the courtyards, but the property managers only collect the money without delivering work.

The property management and maintenance at Juer Hutong new courtyard housing (Case 3) is also inadequate. Although the Residents' Committee (another public entity) assumes some of its maintenance jobs, it is far from enough. As Juer Hutong was redeveloped at the time (1989) when there was no such thing as 'property management' or 'maintenance fees,' it was not a problem then. Right after completion, homeowners paid a property fund deposited to a bank account. However, this fund is prohibited from withdrawals because there are no official property management services. In the beginning, the cost for repairing leaking roofs came from the maintenance fund, but as many other shortfalls such as the garage being occupied by private individuals, are overlooked, the property management exists only nominally, with neither maintenance fees collection, nor services for residents.

However, if all the residents at Juer Hutong were asked again to submit maintenance fees, they would undoubtedly be unwilling to pay, rendering it difficult for the Residents' Committee to collect. At present, the only service available is from the Environmental Protection Group that employs ground sweepers and garbage collectors each day. Meanwhile, the Residents' Committee is planning to establish a 'Homeowners Association' to solve some public maintenance issues (Chair of Residents' Committee, interview, 2008).

Due to a lack of maintenance and management, three of 17 interviewed residents at Juer Hutong new courtyard housing complained that the courtyards have lost their function. At one time, there were plants, flowers, and grass in the courtyards, but they all died because no one watered them. Later, grass in the courtyards was replaced with patio stones. For over 10 years, nobody has taken care of the courtyards; although the trees still stand, nobody maintains them. This finding is consistent with that of Liu Wenjie (1992). However, architect Wu Chen (Professor Wu Liangyong's son) contended that the architect should not be blamed for poor property management because it is not a design issue.

Private Housing Management

Private housing management companies may not perform a better maintenance job if fees are unpaid. At Tongfangyuan (Case 4), one of five interviewed residents noted that the paint on one building is peeling off and giving a rundown image, but the property management advises that it is an expensive paint job, and that the residents have to pay. Two of 14 interviewed residents at Jiaanbieyuan (Case 6) subsequently commented that property management only solves some problems, and that the residents have to find contractors to even unplug toilets.

The initial property management at Shilinyuan (Case 5) was part of the developer's company that always responded to damages and repairs. However, it did not keep a public property account. Some Property Owners Committee members started looking for faults in their management and introduced a new management company that would have to offer benefits to homeowners serving the Committee by exempting their maintenance fees. Consequently, some homeowners have never paid maintenance fees. Although the new management company is proficient with a more professional attitude towards service, its profit margin is so decreased by the Property Owners Committee that it is very difficult to ask them to repair anything for the homeowners if they have to spend money on it. The current maintenance fee at Shilinyuan is lower (CNY ¥0.5/sqm or USD $0.075/sqm) than that in the new Industrial Park District (e.g., CNY ¥2.59/sqm or USD $0.39/sqm) with differences in service quality and sense of responsibility. The differences lie in the size, number, and maintenance of communal green areas and children's playgrounds, supply of health facilities, clear division between car park and pedestrian spaces, organized social events during Spring and Mid-Autumn Festivals, 24-hour duty of property management, professionally trained security guards, tighter security control at the estate gate, and so on.

Car Park Spaces

The Beijing City Planning Committee (2002, p. 14) indicated that the width of Beijing's traditional *hutong* varies from less than 3 m (16%), to 3-5 m (35%), 5-7 m (26%), 7-9 m (17%), and 9 m (6%). Around Xisibei area (Case 1), the widest *hutong* are 9-10 m, the narrowest 3.5 m, and the average width 6 m (Beijing City Planning Committee, 2002, p. 97). Thus, most *hutong* are unsuitable for car traffic or car park. However, residents have nowhere else to park their cars except the *hutong*. Moreover, residents often park their bicycles in the courtyards, making the already limited outdoor spaces even more congested for foot traffic.

**Figure 8.24   The underground communal car park for residents at Nanchizi
new courtyard housing, Beijing. Photo by Donia Zhang 2007**

Planning policies and developers' profit considerations have also affected
parking provisions. The Nanchizi new courtyard housing (Case 2) has about 1000
households, but only 150 parking spaces available in the underground garage. The
architect hoped that residents would use the underground car park to maintain
the 6 m *hutong* width for motor vehicles to drive through in both directions, and
for meeting basic fire safety requirements (Lin Nan, project architect, interview,
2008). However, this parking space provision is obviously inadequate and the
underground parking fee is so high that residents park their cars elsewhere. The
situation worsens when residents from other lanes take over parking spots in the
hutong here, prompting several residents to suggest better security of their spots
using 'parking lot locks.'

Parking design at Juer Hutong new courtyard housing (Case 3) was restricted
by China's economic conditions in late 1980s because there were few private cars
and no parking design standards at the time. The designers did not anticipate such
a fast development of ownership of private cars in China (Liu Wenjie, project
manager, interview, 2008). From the residents' point of view, the shortage of
parking spaces at Juer Hutong has affected the housing estate's overall function.
As the local government sublets the basement to non-residents, they nightly park
their tricycles used for collecting recyclable materials or garbage in the *hutong*,
so it looks messy and untidy. This finding conforms to that of Liang and Zong

Figure 8.25   Private cars parked at a Juer Hutong new courtyard housing estate gate, Beijing. Photo by Donia Zhang 2007

(2005). When cars run on both sides of the *hutong*, walking becomes dangerous for the elderly who are left with no courtyard for sitting or walking, especially in the most recent 2 years (2006-2007) when an increasing number of cars in the area was seen (5/17 respondents). One resident in Courtyard A cynically commented: "Fortunately the courtyard gate is narrow or cars would drive in." As some patio stones outside the courtyard gate are tilted under the weight of cars, some elderly people have tripped and fallen a few times.

Suzhou Tongfangyuan (Case 4) has one garage for about 15 parking spaces per courtyard, which is 15-20 percent parking provision based on the number of households in each apartment building, and each courtyard villa has a private garage. This number is far from enough because there were no design standards for car parks at the time (1996), and the planners did not anticipate the development of private car ownership could be so fast. Residents from apartment buildings can only park their cars in the estate's outdoor spaces or on the roads outside the estate (Qu Weizu, project planner, interview, 2008; Ren Huakun, project architect, interview, 2008). The property manager noted that the concrete road slabs in the estate have tilted under the weight of cars; they are hard for cars to drive on. Since over 10 parking spaces in the underground garage are reserved for shoppers instead of homeowners because the property management receives revenue for

Figure 8.26   Private cars parked on the alley of Jiaanbieyuan, Suzhou. Photo
by Donia Zhang 2007

Figure 8.27   Private cars parked on both sides of the alley where residents
walk or cycle, Jiaanbieyuan, Suzhou. Photo by Donia Zhang 2007

Figure 8.28   Private cars parked on grass at Zhuzhiyuan ('Bamboo Garden Housing Estate,' 2000, a minor case), Suzhou. Photo by Donia Zhang 2007

them, residents complained that outsiders take over their parking spots. This finding is consistent with that of Jin *et al.* (2004).

Due to land shortage, Shilinyuan (Case 5) is provided with bike parking but not a car park (Peng Hongnian, project architect, interview, 2008). A resident complained that too many cars are parked along the estate's alleys, creating disorder. Some residents also honk and speed through the estate, disturbing the peace of the residential environment.

With about 500 households at Jiaanbieyuan (Case 6) and each household having 1-2 cars, the 40 parking spaces in the ground-level garage are certainly inadequate (Property Manager, interview, 2007). The developer later had to convert the property management building into a car park (Liu Weidong, project planner, interview, 2008). Several residents protested that the remaining cars have to park on both sides of the estate's alleys, leaving only one car width for walking, seriously affecting the environment and air quality. During the week of the survey (2007), I observed a severe traffic jam around 5-6 pm every evening when security guards directed a queue of cars at the North Gate.

The Suzhou Urban Planning Bureau requires a new housing estate to provide 100 percent parking spaces and an additional 10 percent visitors' parking (Liu Weidong, interview, 2008; Peng Hongnian, interview, 2008). However, a contradiction exists when insufficient parking spaces impair a housing estate, and too much motor traffic destroys its peacefulness.

Summary and Conclusion

This chapter examined the space and construction quality of the renewed/new courtyard housing in Beijing and Suzhou. The findings suggest that the renewal (Case 1) has not solved the problems of small living space and inadequate basic facilities. While the interior spaces of new courtyard housing units are generally small in Beijing Nanchizi (Case 2) and Juer Hutong (Case 3), they are larger and more satisfactory in Suzhou Shilinyuan (Case 5) and Jiaanbieyuan (Case 6). These results may be related to less restrictive planning regulations but more rigorous construction requirements set by the Suzhou municipal government (Chapter 6).

The findings also indicate that most residents prefer to live in low-rise housing of 1-3 storeys for practical reasons, and living close to the earth (*Feng Shui*) is still preferred. Moreover, 40 percent (n=67) of residents still value traditional Chinese-style furniture for their interiors.

With an obvious lack of construction quality assurance, weak facility provision and poor quality materials and construction have both overshadowed and diminished the reputation of Beijing's new courtyard housing experiments. The Beijing residents' major concern is the deficient functions of the buildings rather than the enjoyment and use of the courtyard. These case projects are in harmony with earth only to a varying extent because the poor quality materials and construction have led to the need for systematic and serious post-occupancy repairs, not to mention excessive waste disposal after extensive renovations. However, in Suzhou, the construction quality issues are much less acute.

Across all the cases, there is a severe lack of parking spaces due to China's economic conditions in the late 1980s when not many private cars were in use at the time of design. This issue may also be linked to land use, as on each plot of land, non-built spaces (such as courtyards) in the middle of the built-up areas count as space and consequently, cars have much less space for parking in these designs. Underground parking is an expensive approach that may not be feasible for modest developments.

In Chapter 9 that follows, social relations among neighbors in the case studies will be investigated, to see whether the renewed/new courtyard housing helps or hinders social cohesion, and what factors contribute to social cohesion.

Table 8.1 Summary of the space of the renewed/new courtyard housing

Spatial Aspects	Beijing Cases			Suzhou Cases		
	1. Renewed traditional courtyard houses	2. Nanchizi new courtyard housing	3. Juer Hutong new courtyard housing	4. Tongfangyuan new courtyard housing	5. Shilinyuan new courtyard housing	6. Jiaanbieyuan new courtyard housing
Interior space	Not resumed to original	45-75 sqm per unit	40-120 sqm per unit	70-95 sqm per unit	90-180 sqm per unit	90-180 sqm per unit
Floor levels	1-storey rooms	2-storey townhouses	2-3-storey apartment buildings	2-3-storey apartment buildings mixed with townhouses	2-3-storey apartment buildings mixed with townhouses	2-4-storey apartment buildings mixed with townhouses
Furniture styles	Varied	Varied	Varied	Varied	Varied	Varied
Facility provision	Deficient	Inefficient	Inefficient	Sufficient	Sufficient	Sufficient
Car park spaces	Deficient	Deficient	Deficient	Deficient	Deficient	Deficient

Source: My summary

Chapter 9

Harmony with Humans: Matters of Social Cohesion in the New Courtyard Housing

Do not do to others what you do not want done to you.
– Confucius, 551-479 BCE, *Analects,* Book 5 Chapter 11; Book 12 Chapter 2; Book 15
Chapter 23

This chapter analyzes the findings related to social cohesion in the renewed/new courtyard housing in Beijing and Suzhou. It explores whether this type of housing facilitates harmonious social relations among humans, which is a fundamental of Chinese philosophy (Chapter 3), and a cornerstone of culturally sustainable architecture in China (Chapter 11). It also attempts to reveal factors that may or may not contribute to social cohesion in Chinese society today.

The chapter investigates four facets of social cohesion within each courtyard housing estate. The first is residents' material conditions measured by their education and house-purchasing power, because socio-economic levels may have an impact on social relations. The second is active social relationships among neighbors through interactions since such communications may offer mutual support, trust, and information that are potential resources for social cohesion. The third is social inclusion or integration of minority groups from different educational or cultural backgrounds into mainstream society, including residents in Beijing's renewed traditional courtyard houses (Case 1) who have lower educational levels and who are relatively poor, and foreign residents who are a minority or marginalized sector of the community in Nanchizi (Case 2) and Juer Hutong (Case 3) new courtyard housing in Beijing. The fourth is indoor-outdoor visual interactions. The study focuses on these four aspects because they are key measurements of social cohesion and are relatively easier to examine in these housing estates.

As such, the social factors studied include: education, occupation, and house-purchasing power, social relations among neighbors, relations with foreign neighbors, and indoor-outdoor visual interactions.

Education, Occupation, and House-Purchasing Power

The survey shows that when combining the percentage of residents with associate degrees, bachelor's degrees, master's degrees, and doctoral degrees, more residents have received higher education in Suzhou (62%; n=123)' than in 'Beijing (46%; n=167), although significantly more residents in Beijing have master's and doctoral degrees (12%) than their Suzhou counterparts (3%). This asymmetry indicates a likely wider education gap among Beijing residents.

The survey also suggests significantly more civil servants and professionals in the Suzhou cases (49%) than those in Beijing (32%), whereas the percentage of laborers (plant and machine operators and assemblers) in Beijing (19%) is remarkably higher than those in Suzhou (5%). This contrast may be due to more residents with higher education in Suzhou (62%) than those in Beijing (46%) and that Suzhou residents enjoy better affordability of their properties.

The study reveals that education *directly* and *indirectly* increases house-purchasing power. Regarding the question, "Has your education helped you find a better job so that buying a house becomes easier?", 63 percent of 40 interviewed residents said 'yes' - education helps them buy a house, but 37 percent said 'no' - their work and field of study do not match, it is their learning ability in higher education that helps with their work success. Five residents admitted that buying a house is the joint effort of spouse or with children, suggesting financial strength in family unity. Four residents felt that China's political milieu in the 1950s and 1960s affected their education, occupation, and house-purchasing power. Work experience equated with formal credentials between the 1950s and 1980s, but it is no longer the case. Economic reforms in China have made education more important in getting better jobs and earning more income than in the past. Five residents maintained that education helps people nurture good social relations conducive to securing employment or higher positions.

However, four of 13 interviewed residents in Beijing's renewed traditional courtyard houses (Case 1) revealed that they cannot afford to buy a new apartment elsewhere in the city but have to remain where they are. This finding confirms a survey report by Zhen Li (2007) that there is a conflict between market housing prices and residents' low income levels in Beijing's old city. The relocated residents displaced by the renewal program have to invest capital to buy new housing units, but most residents cannot afford it.

Social Relations among Neighbors

In *Living in the Old Beijing*, Hequn Bai (2007) recollects his experience of living in a multifamily courtyard compound of Beijing during the 1950s, and perceives that sincerity, honesty, trust, and openness among the neighbors are the most treasured qualities of social relations. This study shows that China's changing and polarizing society has a negative impact on social relations in Beijing and Suzhou. General observations are that neighborly relations in Beijing's traditional courtyard houses (Case 1) have changed over the last 50 years. On average, 10 households share a renewed courtyard. Three of 13 interviewed residents mentioned that people now care only about themselves; there is a big difference between the rich and poor; they are like strangers; and the problem lies in a society that neither emphasizes friendship nor human relations. One resident noted: "Now neighbors are no longer as nice and friendly as 20 years ago [1980s] as people are more conscious of social class, some neighbors are warm but others cold." Socio-economic status appears to affect social inclusion, a finding consistent with that of Yucun Wang (2003).

The survey shows that on average, seven households share a courtyard at Nanchizi new courtyard housing (Case 2). The project architect Lin Nan (2003) wrote that during the sales, many old neighbors still wanted to buy new courtyard housing units in the same compound to continue to be neighbors. However, instead of grouping old neighbors together, Nanchizi new courtyard housing reallocated them according to the number of bedrooms needed by each household, with only 3-bedroom households reallocated in the same compound. After redevelopment, the new neighbors do not know each other, and many do not make an effort to get acquainted. Moreover, due to high housing prices in the inner city, many homeowners let out their units and rent a place further from the city with lower rent to make money from the difference. The survey shows that 69 percent of 71 respondents are homeowners at Nanchizi new courtyard housing who still live there, but 31 percent rent their units to tenants. This arrangement has a significant impact on social relations as residents reported that only old neighbors who know each other visit each other, but new ones who are not acquainted seldom do so (2/16 respondents).

The same situation is observed at Juer Hutong new courtyard housing (Case 3). The survey shows that 57 percent of 56 respondents are homeowners who still live there. The Heating Supply Manager indicated that Juer Hutong has 207 units, of which 40 belong to three organizations; 80-90 are rented out (38%-43%). Juer Hutong thus has a nickname 'United Nations' because of the many different nationalities living there. Two residents revealed that due to socio-economic differences, only old neighbors socialize with one another, but rarely with new neighbors in the courtyard. Professor Chen from Tsinghua University who designed Juer Hutong services and who also lives there revealed that personality affects social relations:

> My apartment has an enclosed private courtyard with no interaction with other neighbors. There is no difference with other apartment buildings. Neighbors whom we did not know are still strangers; we only have contacts with old neighbors. This has much more to do with personality than whether the stairwell is enclosed or not.

Three of 16 interviewed residents at Shilinyuan (Case 5) noted that since residents come from many different places, they bring regional and cultural diversity. Nine residents reported that new neighbors politely say 'hi' when they meet, but rarely visit one another.

Jiaanbieyuan (Case 6) housing estate was once safe and residents did not have to lock their doors at night. However, everyone now securely locks their doors because the social environment is more polarized, with many more thieves and burglars around. When all doors are locked, residents find it inconvenient to go in

**Figure 9.1   Elderly people playing games together in a traditional *hutong*,
Beijing. Photo by Donia Zhang 2007**

and out. Thus, good neighborly relations may be affected by residents' perception
of socio-economic conditions.

Does a courtyard support neighborly relations? The survey results show that
communal courtyards can play a vital role in facilitating social relations. The
majority of survey respondents (64%; n=290) considered the *courtyard* as the
most important space for social relations, with this number higher in Beijing (71%;
n=167) than in Suzhou (54%; n=123), suggesting that Beijing residents use the
courtyard more often as a social space than their Suzhou counterparts. This finding
may be linked to the fact that fewer Beijing residents have a private yard (0.6%;
n=167) than Suzhou ones (38%; n=123). Nevertheless, it also confirms a common
perception in China that northern Chinese are generally more outgoing than their
southern counterparts. My finding is consistent with that of Zhao (2000) that
Beijing's traditional courtyards, even when shared by many families, are where
social interaction takes place, such as playing chess with neighbors, chatting with
friends, or playing with children (pp. 103-105).

While Suzhou residents considered the public corridor (33%; n=123) as their
second most important place for social relations, it is the *hutong* for Beijing
residents (25%; n=167). My result confirms that of Zhao (2000) that although
residents spend less time in the *hutong* than the courtyard, the *hutong* is also a
major venue for the elderly to socialize (pp. 103-105).

Unlike English people who use pubs and markets as popular social spaces (Dines *et al.*, 2006; Watson and Studdert, 2006), only 4 percent of the Beijing residents and 3 percent of the Suzhou residents surveyed regarded their local store/ pub as a setting for enhancing social relations.

Overall, 64 percent of 290 survey respondents got on 'well' or 'very well' with their neighbors, with 31 percent reporting 'OK,' but none got along 'poorly' or 'very poorly' with their neighbors. When compared with residents at Beijing's renewed traditional courtyard houses (Case 1; 50%; n=40) and Juer Hutong new courtyard housing (Case 3; 61%; n=56), residents at Nanchizi new courtyard housing (Case 2; 67%; n=71) coexisted the best with their neighbors. In Suzhou, Jiaanbieyuan (Case 6; 89%; n=44) residents got on better with their neighbors than Shilinyuan residents (Case 5; 48%; n=42), this may be due to the form and space of their estates' design, among other factors.

Renewed Communal Courtyards and Social Relations

After renewal, the same working-class nuclear families have remained on site. However, renewal did not solve problems of small living space, and many communal courtyards have become even smaller and caused disputes among neighbors. Five of 13 interviewed residents in Beijing's renewed traditional courtyard houses (Case 1) observed that their neighborly relations are as good as before, and that they watch out for each other. Nevertheless, *eight* of 13 interviewed residents noted that their neighborly relations are *not* as harmonious as before, and that there is less communication or none at all. This significant finding confirms that of Lu and He (2004).

Residents generally reported that if neighbors were dissatisfied with another before renewal, they could at least tolerate this on the surface to maintain civility. Many households tried to claim more space during renewal by extending their exterior walls, may be just one-brick width, into the courtyard. Right after renewal, the problem became acute with some neighbors having fist fights after tearing off the mask of civility. Subsequently, whoever has a sharp tongue can bully others, and older residents will harass newer ones. Even the police have been called several times to mediate conflict with unsatisfactory outcomes. Neighbors have stopped talking with each other, which is an unprecedented phenomenon (5/13 respondents), followed by a gradual disinterest in resolving their conflict. As the physical construction work is completed, residents generally let go of their grudges. However, a few still hold resentment even if they speak with each other (5/13). Thus, decreased spaces in communal courtyards have had a negative impact on social relations among neighbors.

New Communal Courtyards and Social Relations

As the form and space of a courtyard may impact on social relations, eight of 16 interviewed residents at Nanchizi new courtyard housing (Case 2) commented that their communal courtyards promote communication and contact, and encourage the neighbors to meet as soon as they come out of their units. The neighbors negotiate about such details as where to place belongings and how to plant flowers in the courtyard. To deepen neighborly relations, they sweep the courtyard together, eat there with one another, assist with everyday needs, such as watching out for trespassers when neighbors are away, watering flowers, offering home-cooked food, collecting laundry and mail, helping to access local services, taking a sick neighbor to hospital, and so on. Seven residents commented positively that their social relations with neighbors are as harmonious as in the old courtyard house before redevelopment, and that there is no conflict. Two residents felt that this kind of new courtyard housing is much more conducive to social relations than apartment buildings where people can be neighbors for many years but still do not know each other, including some who share the same workplace.

Three of 17 interviewed residents at Juer Hutong new courtyard housing (Case 3) reported favorably that their communal courtyards increase neighbors' likelihood of personal encounters, and that their homes cultivate social relations more than other housing forms. They also found that their collective home gives a sense of 'traditional courtyard house,' and that the neighbors have better chances to meet as soon as they come out to chat in the courtyard, especially in the summer when they can sit and enjoy the cool air. The neighbors may help each other when in need, while their apartments can offer them privacy. With a full sense of 'human touch,' the neighborly relations are perceived to be harmonious (7/17 respondents).

This finding confirms that of Tsinghua University in 1992 that 60 percent of 31 households said that they know their neighbors primarily through encountering them in the courtyard, and that they enjoy stronger social relations around two larger courtyards than do residents with two small courtyards in the phase one experiment (Wu, 1999, pp. 169-170). Nevertheless, Tsinghua University interviewed five residents 15 years later and found that nearly all the original residents had sold or sublet their units to urban elites or foreigners. Gentrification has gradually occurred due to market pressure and transiency of residents. The new courtyards seldom facilitate neighborly communications because of the changes in social structure, and the insufficient sunlight in the small yards discourages residents to linger (Liang and Zong, 2005). Thus, courtyard size may impact on social relations.

Two of five interviewed residents at Suzhou Tongfangyuan (Case 4) observed good communications among neighbors, and that they chat in a friendly way when meeting in outdoor spaces of the housing estate. However, three residents noted that they exchange 'hello' on the way in and out with superficial knowledge of each other, and that their neighbors will only visit those in the same building, but have no contact with those living further because there is no need or opportunity to do so. This finding is consistent with those of Jin *et al.* (2004) and Chen (2011) who indicate that Tongfangyuan residents comprise mostly business owners and foreign investors: this demographic group is normally busy, and many own properties elsewhere. While many residents stroll at night, walking is not a communal activity. With limited organized activities in the inadequate outdoor spaces, chances are low for residents to get acquainted (Chen, 2011; Jin *et al.*, 2004). Yucun Wang's (2003, p. 23) study even suggests acute conflict among Tongfangyuan residents from different socio-economic levels.

Three of 16 interviewed residents at Shilinyuan (Case 5) enjoyed good neighborly relations, regular visits and chats. For instance, when it rains, they remind each other to bring in their laundry. Another example is of an elderly neighbor with a fracture of the lumbar vertebra. Since her children worked during the day, her neighbor looked after her and provided food that she liked every day until her recovery. For neighbors with closer contacts, they send *zongzi* (a pyramid-shaped dumpling made of glutinous rice wrapped in bamboo or reed leaves) during Dragon Boat Festival and give gifts to each other's children during Spring Festival (2/16 respondents). However, five residents felt that life was more sincere in traditional courtyard houses where the gate was always open with several generations living together, and where the elderly enjoyed visiting neighbors and communicating directly. A resident recalled:

> I lived with over 10 households in a *dazayuan* ['big and mixed-yard'] for a few years. Every time we saw each other when coming and going, we would say 'hi.' We enjoyed cool summer breezes in the courtyard, and had good contact. The kind of housing estate we now live truly hinders our social interaction and communication with neighbors.

Similarly, Jin *et al.* (2004) noted that in traditional Suzhou neighborhoods, almost all the residents living on the same street know each other because when their children visit each other's homes and the elderly chat often, there is a strong sense of social interaction. Newly-built housing estates offer much fewer venues for this cohesion, perhaps because the developers' approach to establishing traditional ambience rely more on architectural applications for exterior features than human factors (Jin *et al.*, 2004).

Four of 14 interviewed residents at Jiaanbieyuan (Case 6) remarked that relations among nearby neighbors are harmonious, there is no quarrel or dispute, and they can negotiate problems that arise. Childbirths and new mothers in their post-partum month of rest will receive eggs dyed red (a Suzhou custom) and blessings from neighbors as newlyweds give out 'candies of joy.' Old neighbors will also send cakes to those newly moved-ins, and many residents exchange gifts or home-cooked New Year food during Spring Festival.

However, five residents at Jiaanbieyuan indicated that multi-storey apartment buildings are not as personal for facilitating warm communication and exchange among the neighbors as traditional courtyard houses or 'big and mixed-yard' compounds. A stronger sense of trust and 'one big family' was present in 'mixed-yard' compounds where neighbors could meet, eat, and chat in the courtyard, which created opportunities to interact and communicate. A respondent reflected: "I still miss the courtyard house from childhood: strong family ties, neighbors' simplicity and honesty were readily demonstrated there. Now that people's living conditions have improved, their opportunities to communicate actually become poorer."

Overall, 12 of 35 interviewed residents at the three Suzhou cases suggested that, compared with traditional courtyard houses up to the 1980s, the *spatial structure* of the new courtyard-garden housing estates is less convenient for neighborly communications, and that it weakens social relations. They indicated that neighborly relations in the new courtyard-garden housing estates are generally less warm, involved, and concerned; neighbors seldom communicate, and there is less affection for one another.

Traditional courtyard socializing usually took place among members of an extended family, and later, among working-class, nuclear families who shared a courtyard compound in the cities. New courtyard compounds are now designed for multiple families, with an independent unit installed with modern facilities for each family. However, social interactions in new courtyards seem different from those in traditional ones. Findings collected from residents in new courtyard housing yielded mixed responses. While people are nostalgic for close relationships as in traditional courtyard houses, modern lifestyles no longer permit such living practices. There are also issues with dog ownership, loud television sets, and noisy air-conditioners that affect the estates' peacefulness (3/82 respondents).

Modern Lifestyles and Social Relations

Fast-paced, modern lifestyles reduce the chances of neighborly communications. When some people work extra hours at night, there is less time for visiting others. Before redevelopment, a family at Nanchizi (Case 2) had a private courtyard house where all their relatives lived together in one compound, but now they live in two new, adjacent courtyard compounds, their relatives visit each other irregularly, because when some come home late at night after a long day's work, most have already gone to bed. A resident explained:

> Busy with my work, I only get home at 8 pm and am limited to chatting with neighbors or gardening in the courtyard on weekends. I am not rich, the pressure from work is heavy, and I do not have much time for activities such as barbecuing in the courtyard.

Twenty-two of 30 interviewed residents at Shilinyuan (Case 5) and Jiaanbieyuan (Case 6) revealed that the neighbors seldom communicate, and that every household lives its own life behind closed doors, mainly because people are very busy with their work. If there is an issue with neighbors, they will try to solve it through the property management. Although they walk around the housing estate and talk with neighbors when they meet, these relationships are superficial and distant. Thus, China's fast-paced economic development extracts a high social cost.

Modern facility provisions such as individual bathrooms and kitchens also reduce the likelihood of neighborly communication. In the past when residents

Figure 9.2   A widened *hutong* that looks toward the Pudusi Temple, Nanchizi new courtyard housing estate, Beijing. Photo by Donia Zhang 2007

Figure 9.3   A sign that reads: 'Parking is prohibited' in a small community park, Nanchizi new courtyard housing estate, Beijing. Photo by Donia Zhang 2007

lived in traditional courtyard houses, they shared one water tap in the courtyard, and a toilet in the *hutong* ('lane'). Neighbors had more opportunities to interact and chat. Now individual homes have self-contained kitchens and bathrooms, neighborly communication and social relations deteriorate despite improved living conditions. For example, at Nanchizi new courtyard housing (Case 2), although the *hutong* is widened, it has lost traditional shops/stores and socialization hubs. Residents sometimes take a walk in the *hutong* or the small community park where they meet neighbours and chat (2/16 respondents).

Social Activities in Courtyards and Gardens

Noise and air pollution generated from social events in communal courtyards is an issue. A resident at Nanchizi new courtyard housing (Case 2) recounted that once, a company organized a tea party in the communal courtyard with a loud and noisy megaphone without the Residents Committee's approval. Such social activities should always apply for approval.

At Juer Hutong new courtyard housing (Case 3), some residents use the communal courtyards for holding parties at night or on weekends with music. An unpleasant incident occurred when a chef who worked for a hotel in Beijing invited a group of friends to a barbeque on his roof terrace. A neighbor reported the

Figure 9.4   Two elderly men playing a game of chess in a Beijing street garden. Photo by Donia Zhang 2007

Figure 9.5   A small group of elderly women sitting and chatting in a Beijing
street garden. Photo by Donia Zhang 2007

Figure 9.6   A group of elderly women dancing in a Beijing street garden.
Photo by Donia Zhang 2007

noise and smoke to the police, but the police let the cookout continue because the meat was already in the cooking process and it would have been a waste to throw it away. Since then, no one has barbequed on roof terraces because the smoke and flames disturb other residents (Chen Zhijie, interview, 2008). A resident at Juer Hutong stated that it is inappropriate to have social activities in residential courtyards because the noise infringes on other people's rest; neighbors have different schedules for retiring and rising, so respect, tolerance, and reciprocity are important qualities for maintaining harmonious social relationships.

Street gardens in Beijing are now important places for social and cultural activities, as they are more public. On a Tuesday morning in early November 2007 and on my way to Juer Hutong new courtyard housing estate from the Gulou (Drum Tower) subway station, I observed two elderly men playing a game of chess on a wooden bench in the street garden along Beijing's northern second ring road. On my way to the Andingmen (Peace Gate) subway station in the evening, a group of (about 10) people were singing old (perhaps 1950s') songs and dancing together. On the Friday morning that week, I saw many elderly people chatting together on wooden stools in the street garden, while a group of women practiced traditional dance.

A communal Central Garden located in the middle of Shilinyuan (Case 5) is paved with patio stones, with grass and bushes on the edges, but no water feature (unusual for a Suzhou garden). A resident explained that because this housing estate was built to meet standards of its time (2000), certain important garden elements are missing. Four of 16 interviewed residents revealed that a handful of elderly residents chat in the Central Garden regularly, but others seldom or never do so; these differences are more related to character, habits, and educational levels than architecture.

On a Saturday afternoon between 2:30-5:00 pm in early October 2007, I observed seven adults sitting on a stone bench in Shilinyuan's Central Garden, chatting and watching children play. Two were young mothers, and five were in their 60s and 70s. One young mother told me that she rarely saw the residents frequent the Central Garden; quite often, only her husband, daughter, and herself would be in the Garden because they did not have their own private courtyard. When her daughter played with other children in the Garden, she could make new friends. Although it was helpful to get acquainted with neighbors through children, mosquitoes were a threat. Another young mother explained that she was in regular contact with the neighbors because of the Central Garden, but she also complained that the Central Garden was too small for the children to bicycle or to roller skate. Although many children often play ball games in the Central Garden, there is no sports ground here. Only schools in the Pingjiang Historic District have sports grounds, but Suzhou schools are closed to the public.

Jiaanbieyuan (Case 6) has a larger communal Central Garden than that of other housing estates in Suzhou's old city. The Garden's landscaping was carefully designed in the style of classical Suzhou Gardens, with a pavilion, fish pond, corridor, moon gates, and bamboos. Two of 14 interviewed residents reported that

Figure 9.7   A group of elderly residents chatting in the communal Central Garden, Shilinyuan, Suzhou. Photo by Donia Zhang 2007

the Garden grows flowers and fruits all year round. Three residents stated that this Garden is advantageous to social and cultural activities where the elderly exercise with others every morning, and where children play under the adults' supervision. A security guard agreed that the Central Garden is frequently used by elderly residents at around 6:30 am every morning, when a group of 20-30 elderly people (mostly women) exercise for 1.5-2 hours. A resident confirmed, "I have morning exercises in the Central Garden at 7:30-8:30 am almost all year round, about 10 of us retirees." Some residents sit or walk in the Central Garden daily except in July and August when it is too hot.

One mid day in mid October 2007, several parents/grandparents were observed taking their children/grandchildren to Jiaanbieyuan's Central Garden, strolling and telling stories. A few children were cycling or skipping (exercising) in the alleys. The property manager and a resident indicated that since the Central Garden is very dark at night, it is seldom used for gatherings. Another resident said that the Central Garden is so small that they prefer to stroll along the canal after dinner.

A Jiaanbieyuan resident complained about being woken up once early in the morning by firecrackers let off by newly-rich neighbors from the countryside, who showed no concern for others, or for norms of civility upheld by most Suzhou citizens. Thus, social activities in communal courtyards or gardens are not shared by all residents, especially when socio-cultural differences are perceived. These dissimilarities may actually cause tension and resentment among neighbors.

Figure 9.8   Parents/grandparents watching their children/grandchildren play in the Central Garden, Jiaanbieyuan, Suzhou. Photo by Donia Zhang 2007

Organized Communal Activities and Social Relations

Organized communal activities may aid social relations. Two of 13 interviewed residents at Beijing's renewed traditional courtyard houses (Case 1) reported having organized communal health activities. Some senior residents listened to health lectures for 2 hours each week at the Community Health Center (Xinjiekou Sanitation and Anti-epidemic Station). A dedicated resident at Qingfeng Hutong also coordinated reading clubs for retirees in Chinese Classics such as Confucian *Analects*, Laozi's *Dao De Jing*, and *Disciples' Rules*. They met in a nearby community garden during warm seasons, and the group had expanded from around 10 people to over 20 in 2007.

Nanchizi (Case 2) Residents' Committee managed various social activities at the Pudusi (Buddhist) Temple where they held weekly classes in weaving, English language, sports (*taiji*, dancing, badminton), science and technology, and spontaneous activities inspired by residents themselves. Four of 16 interviewed residents at Nanchizi attended the activities regularly. A resident revealed her daily routine: "I participate in the community's health activities by going to Jingshan Park in the morning, and the Community Health Club (on Donghuamen Street) in the afternoon where there are aerobics, yoga, and other exercises.... In the

Figure 9.9 Two elderly people talking in Pudusi Temple's courtyard at Nanchizi new courtyard housing estate, Beijing. Photo by Donia Zhang 2007

evening, we all go to the [Pudusi] Temple where we have activities together with neighbors who we know and communicate easily with."

Soon after Juer Hutong new courtyard housing (Case 3) phase one was occupied, the Residents' Committee organized singing and entertaining activities in the communal courtyards. However, no one leads these pastimes now even if they may rehearse perfunctorily for such events as a singing contest for the 2008 Olympics, or other political activities (Chen Zhijie, interview, 2008). The only planned social activity for the elderly is a singing group every Wednesday afternoon in the public activity room at the Community Center. Two residents have proposed plenty of exercise equipment, a ground for ball games, a swimming pool, and other such facilities. Still, some residents managed to exercise at a district sports center where they found skating, yoga, and other fitness programs.

Two of five interviewed residents at Tongfangyuan (Case 4) indicated that the property management or Residents' Committee does not arrange social activities, and that organized communal activities are missing in this housing estate. This finding conforms to that of Jin *et al.* (2004). Some residents wished to see planned social events happen in the near future. Nonetheless, one retired resident who became the producer and director of the district community's cultural and artistic subgroup participated in such local arts as singing, dancing, writing and producing traditional plays, and performing in operas.

Two of 16 interviewed residents at Shilinyuan (Case 5) complained that the district community to which Shilinyuan belongs has incomplete facilities; there is no library, reading room, or room for residents to meet in. Some residents exercised in the district Community Center's gym, 5 km away. Two residents suggested that social and cultural activities should be organized by the Residents' Committee or the property management. However, few activities are currently (2007) arranged that residents would enjoy and benefit from.

When homeowners purchased their housing units at Jiaanbieyuan (Case 6), the developer promised to provide them with a community club. However, after the buildings were completed, the club house was rented out to an outside organization. An outdoor swimming pool on top of the garage building has sat completely unused since the SARS epidemic (2003), and is very hot even for swimmers in the summer. As the estate tennis court charges a fee, it is neither used much nor well maintained (Property Manager, interview, 2007). Nevertheless, one of 14 interviewed residents managed to attend 1.5-2 hour classes three times a week at the Seniors' College held at the Canglang District Community Center where classmates socialize after class. An aging population is a major phenomenon in China; social and cultural activities in the communities may benefit everyone's health and wellbeing, especially that of the elderly. Several residents suggested organized meetings in the communal courtyards to counteract social isolation and to improve communication among the residents.

Relations with Foreign Neighbors

Suppose we try to locate the cause of disorder,
we shall find it lies in the want of mutual love.
But what is the way of universal love and mutual aid?
It is to regard the state of others as one's own,
the houses of others as one's own,
the persons of others as one's self...
If everyone in the world will love universally;
states not attacking one another;
houses not disturbing one another;
thieves and robbers becoming extinct...
                                    – Mozi, c.470-391 BCE, Book 4 on *Universal Love*

China has never been a monocultural country; currently 56 different ethnicities coexist under one political regime. From the past to the present, foreign cultural influences have always played a part in forming what is now widely acknowledged as the prevailing Han Chinese ethnic group. Buddhism from India (or Western Asia) in particular, and Islam from Persia (or Central Asia) to a lesser extent, have melted into the local Chinese culture (Ching, 1993; Kohn, 2008; Qiao, 2002; Schwartz, 1985). In recent decades, globalization and China's rapid economic development have attracted an influx of foreigners in major Chinese cities where they work, study, and some settle down permanently. Beijing has become an international capital in that over 100,000 expatriates from over 50 countries reside there (King, 2004, p. 119). An acculturation is happening in China that brings diversity and richness to the existing Chinese culture, but it also demands a higher degree of social cohesion.

Communal Courtyards and Relations with Foreign Neighbors

In a new courtyard compound at Nanchizi (Case 2), five of seven households are foreign families. Chinese residents indicated that most foreign residents speak good Chinese, and that they also have a good understanding of a 'sense of human touch.' Foreign residents express gratitude when a Chinese neighbor helps to collect their laundry in the courtyard from the rain, or when they water their plants. Occasionally, Chinese and foreign neighbors eat and chat harmoniously under grape trellises. Even though a German doctoral student moved out due to an increase in rent, she returned to visit her Chinese neighbor several times.

Likewise, at Nanchizi new courtyard housing, an English scholar and lecturer in the history of Chinese medicine at the University of London revealed that she likes the courtyard and finds it very helpful. Despite the typical Chinese suspicion of foreigners, she gets a very pleasant response as soon as she speaks in Chinese, and her neighbors are always ready to lend a hand. She is especially delighted in watching the baby next door from his mother's pregnancy to his toddlerhood. The courtyard calms, isolates, and connects her at the same time. When she is in London, she dreams of being at Nanchizi where she can develop an effective routine.

A French businessman indicated that except for language barriers, there are no major problems when socializing with Chinese neighbors. The communal courtyard definitely facilitates knowledge exchange among neighbors who are usually friendly, who share many little intercultural activities, and who mutually help with picking up mail, leaving messages, and other daily routines.

As Juer Hutong new courtyard housing (Case 3) won the *World Habitat Award* in 1992, many foreigners like to live there for its reputation, proximity to cultural streets (*Gulou dong dajie, Nanluogu xiang*) and ancient relics (Bell and Drum Towers), and for learning local customs through Chinese neighbors, as an old Chinese saying goes: "When entering a village, do as the villagers do" (入乡随俗, similar to the English idiom "When in Rome, do as the Romans do"). About 40 of 207 units (20%) were rented to foreigners, with 16 of them participating in the survey (2007).

Several foreign residents at Juer Hutong commented that this new courtyard housing facilitates social interactions among neighbors more than a 'modern' Western-style apartment building because of its form. The neighbors frequently come across each other in the courtyard, see one another on balconies or roof terraces, and subsequently make friends. Nevertheless, the courtyards do not have benches or chairs. An American resident observed: "The courtyard and stairwells are where social interactions happen as I frequently meet people there. If it is a weekend or an evening, people are often outside and occasionally chatty. On occasions, the courtyard structure has allowed me to have social gatherings larger than what my apartment can hold." Thus, the new, communal courtyards at both Nanchizi and Juer Hutong appear to have positively supported social interaction and relations with foreign neighbors.

Language Barriers and Cultural Differences

Different foreign residents have diverse experiences in Beijing's new courtyard housing. At a new communal courtyard in Nanchizi (Case 2), hardly any interaction takes place between Chinese and foreign residents because of language barriers and cultural differences, except that foreign residents sometimes ask their Chinese neighbors about bill payment. A similar situation occurs at Juer Hutong (Case 3) where Chinese residents seldom communicate with their foreign neighbors, except with those who can speak some Chinese, and where both groups will just say 'hi' when seeing each other in the courtyard. When they need to pay bills, some foreign residents will ask their Chinese neighbors about maintenance issues in the absence of property management.

A German-Montenegrin doctoral student who can speak and write proficient Chinese commented that when the communal courtyard at Nanchizi (Case 2) is frequently used by neighbors' outdoor activities, it is not an appropriate space for private reading or wine-drinking. As all outdoor (and indoor) activities are registered and even monitored by Chinese neighbors, this scrutiny is an infringement on one's privacy. She stated that 4 years of interaction with good or bad neighbors has been enough for her, and that closer proximity may cause many problems. She also noted that Nanchizi was almost all Chinese occupied, but it has now 'advanced' to yet another foreign domain in Beijing that sometimes leaves her out of touch with China. Similarly, two Chinese residents at Nanchizi new courtyard housing complained about too many foreigners living there, affecting China's image.

An Italian-French resident at Juer Hutong (Case 3) indicated that the new communal courtyard is not as conducive to social interaction as a traditional one because too many households are sharing it. He contended:

> I don't think my courtyard is particularly designed for socialization. A good friend lives a few doors away in a traditional courtyard in Juer Hutong with only four families, and her interaction with neighbors is much better. I usually talk with only one lady who always collects garbage outside and is the 'guard' of both the courtyard and Beijing's memory. But there are too many people and too many buildings within the new courtyard to facilitate any real social connections.

These comments suggest that high population density negatively impacts on the use of the new communal courtyard.

Different Lifestyles and Communal Courtyards

A major problem between Chinese and foreign residents is lifestyle differences. Several Chinese residents said that foreign residents at Nanchizi (Case 2) often hold parties in the new communal courtyards, inviting only foreign friends. They burn candles and make a mess with leftovers in the courtyard; they are also noisy and disorderly, which irritates Chinese neighbors. However, from some foreign residents' point of view, the communal courtyard is where social interaction should take place.

Many foreign residents at Juer Hutong new courtyard housing (Case 3) enjoy night life that their Chinese neighbors cannot get used to. For example, some foreign residents only come home at 2-3 o'clock at night. They wake up other residents when they climb the stairs as these buildings' public areas have poor sound insulation.

Five of 17 interviewed residents at Juer Hutong new courtyard housing stated that only foreign residents like to hold parties in the courtyard on weekends with mainly foreign guests, and that only Chinese residents who speak English or French well may be included. They light candles and play musical instruments that can be messy and noisy. An unpleasant incident occurred when a high-school student in a Chinese family was preparing for his term exams, but noise from a party was so loud that he could not concentrate on his studies and subsequently called the police. Although the party host apologized when the police came, the party resumed afterwards. Thus, mutual consideration and understanding is needed between Chinese and foreign neighbors.

Chinese Neighbors in the View of Some Foreign Residents

In the eyes of some foreign residents at both Nanchizi (Case 2) and Juer Hutong (Case 3) new courtyard housing, their Chinese neighbors are not very sociable because they rarely hold parties in communal courtyards. In their view, communal courtyards are appropriate places for social activities. An American resident at Juer Hutong said, "My neighbors and I are cordial but not overly social. I have been invited once or twice to social events by them, both times by non-Chinese residents. My Chinese neighbors sometimes smile and say 'hi,' but rarely more." Thus, language barriers, differences in cultural backgrounds and lifestyles may affect the use of communal courtyards.

Indoor-Outdoor Visual Interactions

Indoor-outdoor visual interaction through windows is a unique feature of courtyard housing. Adults can supervise children's activities in the courtyard while doing something else at the same time. While 90 percent of 290 survey respondents have windows facing a courtyard, 54 percent watch and take care of children's play in the courtyard through their windows daily, and 37 percent weekly. In general, residents spend an average of 30 minutes each time watching children play in the courtyard.

Indoor-outdoor visual interaction through windows is a positive experience for some residents at Nanchizi new courtyard housing (Case 2). A resident revealed:

> I can continually keep my private and public space. The big window at the front gives me ample light and a connection with the courtyard, such as watching the boy next door grow up…I often speak with neighbors through the 1st-floor window. I even keep the door open and I feel safe because I know my neighbors are in the courtyard.

At Juer Hutong new courtyard housing (Case 3), four of 17 interviewed residents shared their positive experience of indoor-outdoor visual interaction in the communal courtyard. For example, a resident watched the children play shuttlecock or badminton in the courtyard, and kept the door open all summer for fresh air and children's voices. Although each household is independent in their apartment, communal courtyards give neighbors visual connections so that they share a sense of belonging to the courtyard (3/17 respondents).

Another resident at Juer Hutong new courtyard housing recounted that when her grandson was little, he played with five or six other children in the courtyard while some elderly people watched them either in the yard or through their windows at home. The children were noisy and would disturb other neighbors who wanted to rest. Now the children have grown up, they seldom play in the courtyard. Yet another resident revealed that some children use the courtyard for activities on weekends, and some children from nearby neighborhoods will also play in the courtyard during summer vacations because their parents think it is the only open space available in the area that is safe, unlike the streets full of cars. However, these activities can also cause damage, such as once when children used the gate eaves as a basket ball net, they smashed them with the ball.

A resident at Tongfangyuan (Case 4) observed that while the idea of having a kindergarten in the housing estate is helpful, teachers taking children out for a lunchtime walk may disturb some residents.

Thus, while some residents have positive experiences with indoor-outdoor visual interactions and children's recreation in the communal courtyards, others may not concur.

Summary and Conclusion

This chapter explored the social cohesion among neighbors in the renewed/new courtyard housing in Beijing and Suzhou, to see whether these housing designs help or hinder social relations, and what factors may or may not contribute to social cohesion. The findings suggest that the renewal of Beijing's traditional courtyard houses (Case 1) have caused some unpleasant disputes among neighbors when they try to increase their interior space in the communal courtyard. This behavior contradicts the doctrine of 'harmony with humans' as emphasized by Confucian ethics. Perhaps because of the form and space designs, Nanchizi new courtyard housing (Case 2) and the larger courtyards at Juer Hutong new courtyard housing (Case 3) generally facilitate better neighborly communication and social relations than the three Suzhou cases.

Although some new communal courtyards support neighborly communication and social relations, the findings are sometimes contradictory, indicating complexity. Their effect is not as noticeable as in traditional courtyard houses of the past. Several residents still hold fond memories of kind neighborly relations in multi-household courtyard compounds between the 1950s and 1980s. Residents' recollections show that redevelopment and reallocation have destroyed much of their neighborhoods' original structure, and that gentrification has significantly weakened neighborly communication.

The results also suggest that neighborly relations are partly influenced by the form and space of the courtyard housing, and partly by a changing and polarizing society, socio-economic differences, housing tenure, modern lifestyles, community involvement, common language, cultural awareness, and the cultural background of the residents.

In Chapter 10 that follows, residents' cultural beliefs and behavioral patterns in the renewed/new courtyard housing will be investigated, to further explore what factors help or hinder their use of the courtyard.

Chapter 10

Harmony with Self: Time and Cultural Activities in the New Courtyard Housing

Heaven in its motion gives the idea of strength.
The wise man/woman, in accordance with this,
nerves him-/her-self to ceaseless activity.

– Yi Jing, the First Hexagram, the Great Symbolism

This chapter analyzes the findings on the time and cultural activities in the renewed/new courtyard housing in Beijing and Suzhou. It explores the cultural beliefs and behavioral patterns of residents and the space they use for their activities, to see whether the housing facilitates a harmonious relationship for residents in terms of their healthy personal development, which is a fundamental of Chinese philosophy (Chapter 3), and a cornerstone of culturally sustainable architecture in China (Chapter 11). The cultural beliefs and behavioral patterns studied include residents' philosophy and religion, their cultural activities and festivities at home, in the courtyard, balcony, roof terrace, community center, community or city parks or gardens. Residents' celebrations of birthdays, wedding ceremonies, and anniversaries are also observed.

Philosophy and Religion

Communism has made a major impact on contemporary Chinese people's belief system. Regarding the question, "Do you believe in any philosophy or religion? Is this belief reflected in your home decoration?", 77 percent of 82 interviewed residents had *no* philosophy or religion. Professor Chen from Tsinghua University recalled that having any other philosophical belief than Communism was illegal in China during the Cultural Revolution (1966-1976); even the *93 Learning Society* (a league of intellectuals established in 1946) was banned, let alone religions (interview, 2008). Five percent (5%) of interviewed residents replied that they are Communist Party members; they have faith in the Communist Party, which does not allow them to believe in any other faith. A resident revealed: "Before retirement, my work unit sponsored me to go to the Sparetime University for special training in adult education using television. I enjoyed philosophy and studied materialism. I believe in the Communist Party that provides me with money." Some even claimed, "I believe in Mao. He was a deity whose bronze statue is placed in our living room."

Several other residents indicated that since they are in natural or medical science fields they are not interested in the supernatural, as a resident explained:

> I am a Chinese Muslim (Huimin), but I am not religious. I have not had much contact with philosophy or religion as I received a formal national education in natural sciences. I regard philosophy and religion as cultural phenomena or cultural knowledge. When we travel, we may visit Buddhist temples without believing in them.

Thus, it may be deduced that most contemporary Chinese people are secular, and that agnosticism or atheism is a major phenomenon in China.

Yi Jing and Feng Shui Theories in Contemporary China

A few interviewed residents were interested in *Yi Jing* (The Book of Changes). For example, a resident at Juer Hutong new courtyard housing (Case 3) revealed that her father studies *Yi Jing* and their apartment has been chosen to suit his ideas and tastes, with Chinese ink wash paintings of landscape and '虚' (*xu*, meaning 'void' or 'emptiness') to reflect Daoist philosophical worldview.

Regarding *Feng Shui* theory, a resident at Juer Hutong new courtyard housing commented:

> According to *Feng Shui*, this courtyard was not well designed because only one tree in the courtyard resembles '困' (*kun*, the Chinese character composed of wood '木' inside a square '口' to mean 'being held'). There should be at least 2-3 trees in the courtyard. Also, they should not be pine or cypress as only graveyards in old China would have these two trees.

A resident at Shilinyuan (Case 5) also explained:

> I make reference to *Feng Shui* theory and I believe some of it. We live on the ground/1st and 2nd floors with the back door facing the staircases. As this is ominous for wealth according to *Feng Shui*, we thus altered this door by sealing it and using the courtyard door instead.

A resident at Jiaanbieyuan (Case 6) reflected: "When the basic requirements are satisfied, I will attend to *Feng Shui*, for example, moving the kitchen stove and the bathroom door's opening direction."

A foreign (French) resident at Nanchizi new courtyard housing (Case 2) also said: "If we buy another house, we will build it according to *Feng Shui*. This one is not because the kitchen and bathroom are not in the 'right locations.'"

Feng Shui theory and practice seem to be growing interests in Beijing and Suzhou, a finding that confirms those of Ch'ü (1994), Knapp (2005c), and Weller and Bol (1998, p. 332).

Chinese Philosophy in Contemporary China

Despite the Communist Party's past attempt to eliminate traditional Chinese philosophy and to convert its population to Communism, a number of residents still hold deep-rooted traditional faith. Six percent (6%) of 82 interviewed residents showed interest in learning Confucianism, Daoism, Buddhism, and Neo-Confucianism. After all, these ideologies have existed in China for several millennia. For example, a resident said, "I like them all – Confucianism, Daoism, and Buddhism, and I respect all spirits and gods. Although I am not devout, I believe very much in them." Another resident revealed, "Although I have not read traditional Chinese philosophy, I have been influenced by Chinese cultural sayings." Several other residents also reported that although Chinese philosophy is not plainly reflected in their home decorations, their hearts and behavior follow traditional doctrines to do kind, morally right things and to be forthright, as a resident elaborated:

> I am interested in Confucianism, and philosophies of Laozi and Zhuangzi. Confucian righteousness (not to act against one's conscience), Laozi's 'harmony with nature,' and one's happiness stems from one's contentment with what one has, all the above concepts are correct. Our home decoration is very simple, unadorned, and conforms to Confucian and Daoist principles.

Another resident answered: "My life philosophy is to be detached from seeking fame or wealth, and being free from material trappings," reflecting Daoist philosophical influence. Yet another resident suggested: "Have an open mind, happy spirit, and you will have a healthy body," indicating a holistic belief in the body-mind-spirit connection.

Seventeen percent (17%) of 82 interviewed residents revealed that they choose to follow Buddhist doctrines, although 5 percent neither join a Buddhist organization nor participate in Buddhist activities. Seven percent (7%) have a Buddha statue placed at home and sometimes burn incense to Buddha, or reserve a quiet space at home for meditation. Five percent (5%) visit Buddhist temples regularly during Spring Festival, and on the 1st and 15th days of each month.

Although Christianity was introduced to China by Western missionaries as early as the 7th century, it has never gained a vast popularity (Ching, 1993; Lutz, 1971), due to widespread and pervasive Confucianism. Two of 82 interviewed Chinese residents believe in Christianity. One resident converted to Catholicism because of a personal influence from Taiwan. Another resident reported her belief in Jesus Christ without participating in any Christian activities.

Cultural Activities at Home

The case studies show that modern lifestyles have largely changed residents' behavioral patterns. When asked about their focus at home, 60 percent of 290 survey respondents reported watching TV, followed by 41 percent using the computer for work or recreation, 28 percent spending time with children, and 23 percent mostly sitting around the dining table. Suzhou residents spend more time on their children (35%; n=123) than Beijing ones (22%; n=167).

Out of 82 interviewed residents, the major activities in an ordinary day at home include: watching TV/movies (37%), using the computer (30%), reading books/ newspapers (24%), cooking (22%), cleaning and tidying the house (17%), eating and drinking (13%), sleeping and resting (12%), chatting/socializing (12%), caring for family members (11%), exercising (7%), writing (6%), practising calligraphy (2%), doing laundry (2%), watering flowers (2%), feeding pets (2%), sewing (1%), playing games (1%), listening to the radio (1%), and carving seals (1%). Thus, some traditional cultural activities, such as practising calligraphy and sewing, are now undertaken infrequently at home, likely due to modern lifestyles with little leisure time for working individuals.

Sleep and work seem to take up a resident's longest time each day. The time diary results reveal the main activities at home as sleeping and napping (average 8.38 hours), working (at home) (7.25 hours), playing games (computer, chess, *majiang*, cards) (3.63 hours), studying (reading, writing, drawing) (2.78 hours), chatting/socializing (2.3 hours), browsing the Internet (2.06 hours), and watching TV/movies (1.88 hours).

Surprisingly notable is that playing games takes up the longest hours in a day next to sleep and work, but only 1 percent of residents mentioned it as a main activity at home, which may suggest that people do not perceive play as a serious activity. Nevertheless, the survey confirms that playing games is the fourth major activity in courtyards (Section 3). Whereas watching TV/movies is their first major activity at home (37%; n=290), and this finding conforms to that of Lu and He (2004), it actually occupies the least number of hours (1.88) among the activities that residents have mentioned. What they learn every day from watching television possibly has a bigger impact on their knowledge and life, and thus gives them a deeper impression. Hence, it seems not the number of hours spent on doing an activity, but the contents of it that matter more to residents. However, due to the small sample size of 22 time diary participants, the average hours shown above is only indicative.

Living Room

The living room (45%; 4.14 hours) is used most often for cultural activities indoors, followed by the bedroom (28%; 10.75 hours), study room (21%; 4.04 hours), kitchen (5%; 1.71 hours), dining room (2%; 1.21 hours), and bathroom (2%; 0.53 hours). These findings suggest that residents spend the longest time in their bedrooms (some residents at Nanchizi new courtyard housing use their bedrooms as office during the day), although sleeping does not require much space. Thus, the living room is the main activity room, and the dining room is used less than the kitchen.

Another study by Yoon and Choi (2001) finds that in contrast to traditional Chinese courtyard houses, the Juer Hutong new courtyard housing has discarded spatial gender division. The kitchen, once reserved for women, is now open to both genders. This change demonstrates a shift in contemporary Chinese lifestyles and domestic roles.

Cultural Activities in Courtyards

The multi-household traditional courtyards could satisfy cultural needs of Chinese families when the courtyard served as extension of the homes, playground for the children, and space for activities such as growing plants and flowers, looking after birds or fish, playing table games, cooking, storing coal and vegetables, parking and repairing bicycles, laundering and drying clothes, getting sunshine, exercising, and so on (Ekblad and Werne, 1990a, 1990b; Zhao, 2000, pp. 103-105).

This research findings show that the top five cultural activities in the renewed/new courtyards are (1) maintaining health/natural healing (mean=3.25 on a scale of 1-5; n=167; average=1.87), (2) drinking tea (m=3.27; n=147; a=1.66), (3) gardening (m=3.29; n=139; a=1.58), (4) playing games (m=3.02; n=108; a=1.12), and (5) holding birthday parties (m=3.2; n=83; a=0.92). This order is confirmed by a further question concerning their *favorite* cultural activities in the courtyard. Thus, maintaining health/natural healing is a courtyard's primary function in the case studies.

Nanchizi new courtyard housing (Case 2) is generally observed to have higher mean and average values in cultural activities than the other two Beijing cases; these new courtyards may be more accommodating. Despite close building distances of 7-9 m at Nanchizi and some residents raising a privacy concern (Chapter 7), five of 16 interviewed residents commented positively that the communal courtyard offers pleasant space for cultural activities. Elderly residents can walk around the courtyard while caring for family members, talking and sharing a drink with neighbors, growing a vineyard, planting flowers, watering plants, raising pets, exercising (e.g., *taiji*), eating, reading, learning a foreign language, and so on.

A resident at Suzhou Tongfangyuan (Case 4) organized cultural activities for neighbors to get together such as a barbecue, and bicycle lessons for children inside and near the housing estate.

In the private courtyards at Shilinyuan (Case 5), six of 16 interviewed residents mentioned that they exercise, hang out laundry, and plant grass and flowers (sweet-scented osmanthus, wintersweet, azalea, clivia, etc.) because the courtyards have a water tap for washing and watering plants. Several residents do *taiji* between the apartment buildings, or enjoy gardening, reading, drinking tea, and resting on roof terraces in the summer. Loft spaces are also well used activity areas for the residents.

A resident at Jiaanbieyuan (Case 6) revealed: "My private courtyard of 45 sqm is a piece of sky and land belonging to us and highly utilized. After dinner, we stroll in it." Another resident noted: "Compared with single-storey traditional courtyard houses, multi-storey apartment buildings undoubtedly prevent daily activities. When we stepped outside the door in the past, we had nature. Living on the 4th floor now, we have lost contact with nature." Yet, a balcony is reported to be useful for growing potted plants (2/14 respondents).

The survey of 290 respondents shows which cultural activities are *not* practised much in courtyards: composing poetry or essays (m=2.3 on a scale of 1-5; n=71; a=0.56), practicing calligraphy (m=2.19; n=72; a=0.54), traditional dancing (m=2.04; n=75; a=0.53), playing musical instruments (m=2.24; n=67; a=0.52), painting (m=2.03; n=65; a=0.46), holding wedding ceremonies (m=2.02; n=61; a=0.42), or singing Chinese opera (m=1.82; n=65; a=0.41). Modern lifestyles possibly no longer allow such activities to be cultivated.

Many homeowners at Nanchizi (Case 2), Juer Hutong (Case 3), and Tongfangyuan (Case 4) rented out their units. With renters mostly working or studying during the day, not many people, and certainly even fewer activities are seen in the courtyards. A resident at Shilinyuan (Case 5) insightfully observed that Suzhouers used traditional, communal courtyards more often in the 1970s because without television, computer, Internet, air-conditioner, or electric fans, people were out more often than now to rely on natural cooling. Hence, not only does design matter to people's activities, modern lifestyles have also changed their behavioral patterns.

Time and Cultural Activities in Courtyards

Time is a constraint to cultural activities in courtyards. The survey of 290 residents shows that they normally undertake their *favorite* cultural activities in courtyards (m=2.65 on a scale of 1-5; n=219) after work: afternoon to evening 4-8 pm as their most active time (41%), followed by morning 8-12 noon (28%), afternoon 12-4 pm (22%), night 8-12 (11%), and dawn 4-8 am (9%). Most residents lingered for 1-2 hours in their courtyards (54%), while others spent 0-1 hours (26%), or even extended this time to 3-4 hours (18%).

Cultural activities that residents spent the longest time on in courtyards are holding birthday parties (4.5 hours), socializing/chatting (2.92 hours), drinking tea (2.5 hours), playing games (1.75 hours), studying (reading, writing, drawing), gardening, and playing with children (1.5 hours), playing musical instruments (1.2 hours), and exercising (walking, yoga, *taiji*, badminton, table tennis, boxing, etc.) (1.16 hours). Thus, a courtyard is an important space for these cultural activities.

Moreover, cultural activities in courtyards mostly take place on weekends (m=2.71; n=219), with Saturdays (46%), Sundays (39%), and Fridays (21%) being the most active days, but Mondays through Thursdays are mainly working days when the courtyards are less used (15%-16%).

Despite time constraint, most residents (53%; n=201) used their courtyards at least 1-3 times weekly, with others using them at least 1-3 times monthly (31%). Nanchizi new courtyard housing (Case 2) had the highest percentage of residents (38%; n=71) using the courtyard at least 1-3 times weekly, compared with that of Beijing's renewed traditional courtyard houses (25%; n=40; Case 1) and Juer Hutong new courtyard housing (23%; n=56; Case 3). In Suzhou, more residents at Jiaanbieyuan (Case 6) used their courtyards at least 1-3 times weekly (59%; n=44) than those of Shilinyuan (43%; n=42; Case 5). These findings confirm on-site observations that the new courtyards at Nanchizi and Jiaanbieyuan are better landscaped and maintained than the other cases.

Climate and Cultural Activities in Courtyards

Climate seems to be another compelling factor impacting on residents' cultural activities in courtyards. Suzhou residents (n=123) are generally noted to have a significantly higher degree of use of courtyards for cultural activities than their Beijing counterparts (n=167) when comparing their mean and average values and percentages of time of the day, length of time, day of the week, frequency, and seasons they spent in courtyards. One reason appears that Suzhou has a warmer climate (average annual temperature 15°-17°c) than that of Beijing (average annual temperature 10°-12°c). Nevertheless, Beijing's annual frost-free period is between 180-200 days (China.org.cn, n.d.), with abundant sunshine during sunlight hours, arguably suitable for outdoor activities.

For Beijing residents (m=2.5 on a scale of 1-5; n=114), summer (53%) is their most active season, although 7 percent of 82 interviewed residents disliked lingering in the courtyard because of mosquitoes and strong, direct sun exposure. Spring (47%) and autumn (47%) are the second most active seasons as they are serene, and winter (13%) is their least active season due to severe, cold days. For Suzhou residents (m=2.94; n=102), autumn (70%) and spring (68%) are their most active seasons whereas summer (33%) is less so because of the heat. There is no dramatic decline in Suzhouers' winter cultural activities (31%) because their city's coldest season is relatively mild.

Courtyard Ownership and Cultural Activities

Only 0.6 percent of 167 Beijing survey respondents had a private courtyard, whereas 38 percent of their 123 Suzhou counterparts had a private courtyard. This finding may explain the higher degree of cultural activities conducted by Suzhou residents. Four of 33 interviewed residents at Nanchizi (Case 2), Juer Hutong (Case 3), and Tongfangyuan (Case 4) revealed that they do not use the courtyard for cultural activities because it is not their private space.

The provisions of balconies and roof terraces for upper-floor apartments at Juer Hutong new courtyard housing have considerably reduced the pressure to use communal courtyards on the ground/1st floor (Wu, 1999, pp. 165-169). When asked where they would conduct cultural activities if not in courtyards, 12 percent of 290 survey respondents reported their balconies or roof terraces. The Juer Hutong residents indicated that the south-facing balconies receive such abundant sunshine that some balconies are equipped with an umbrella. Three of 17 interviewed residents there mentioned that they often have breakfast and tea on the balcony in the summer; it is also a good place to read in summer days and evenings.

Courtyard Size and Cultural Activities

While classical Chinese courtyard houses could be used for a variety of cultural activities, 22 of 40 survey respondents (55%) in Beijing's renewed traditional courtyard houses (Case 1) disclosed that little courtyard space is left. A resident noted that none of the courtyards in the entire *hutong* has any outdoor space left. Moreover, too many bikes are parked in the courtyards after work hours, obstructing residents' movement. Eight of 13 interviewed residents indicated that they would *love* to undertake cultural activities in the courtyards if they were bigger, and that there should be more consideration for outdoor safety.

A resident at Juer Hutong new courtyard housing (Case 3) explained that his courtyard was originally 40 sqm where his family could raise fish in it. To solve the problem of small living space, he added a room in the courtyard, reducing it to 20 sqm, which can only be used for storing flowerpots, not for activities. Four of 17 interviewed residents at Juer Hutong also complained that because their communal courtyard is small and cramped by erratically parked bicycles and amid hung laundry, they do not enjoy much sunshine in it and find it unpleasant to partake in any cultural activities.

Two of five interviewed residents at Tongfangyuan (Case 4) also commented that their private courtyards are too small for cultural activities. This finding confirms that of Jin *et al.* (2004).

Private courtyards of the town/terraced houses at Shilinyuan (Case 5) are normally 20 sqm, with slightly larger ones at 30-40 sqm. Two of 16 interviewed residents expressed that although some courtyards are landscaped with a pond, flowerbeds, grass, and concrete paving, their use of them for activities is low due to the small spaces; the yards are mainly open views that residents enjoy from their windows. For apartment buildings, the front yard is as small as 3 sqm and only useful for drying laundry. To build more units, the developer has made courtyards much smaller than they should be, but this limited feature does not function well for residents. Thus, a courtyard size impacts on residents' cultural activities.

Courtyard Facilities and Cultural Activities

Residents at both Juer Hutong (Case 3) and Shilinyuan (Case 5) complained about a lack of public or recreational facilities in the courtyards for cultural activities. At Juer Hutong, most new courtyard areas are paved with patio stones, although a few types of plants or trees may be present. Six of 17 interviewed residents at

Juer Hutong commented that their courtyards lack benches and seats. During my visit at 2-3 pm on a Thursday afternoon in September 2007, no one was using the courtyards except two to three elderly women sitting by a gate chatting and enjoying the sun. If stone tables and stools were placed in the courtyards, elderly residents could sit, sip tea, or play games, although these activities may also generate unwanted noise.

Cultural Activities in Community/City Parks/Gardens

As community/city parks/gardens are more public than courtyards, when asked where they partake in cultural activities if not in courtyards, 49 percent of 290 survey respondents answered the community/city parks/gardens, and 14 percent said the *hutong/xiang/nong* (lanes/alleys). Because significantly more Beijing residents (75%; n=167) spent their time in public outdoor spaces than their Suzhou counterparts (47%; n=123), regional cultural differences seem to have some effect on the use of public outdoor spaces.

Three of 13 interviewed residents in Beijing's renewed traditional courtyard houses (Case 1) strolled or exercised in Beihai, Jingshan, or Yuyuantan parks in the morning because their living environments offer them too little space for these activities. This finding confirms that of Lu and He (2004).

Two of 16 interviewed residents at Nanchizi new courtyard housing (Case 2) explained that because of the poor landscaping in their community park, they normally just walk their dogs and do little else there. Fortunately, there are sufficient cultural and recreational places near Nanchizi. Two residents indicated that they often walk in the Cultural Palace just east of the Forbidden City, three others reported that they usually stroll inside the Forbidden City and Jingshan Park, and another two revealed that they like to stride in Changpuhe Park ('Calamus River Park' or 'Iris River Park,' built in 2002) behind the red imperial city walls near Nanchizi new courtyard housing estate.

Three of 17 interviewed residents at Juer Hutong new courtyard housing (Case 3) walk or exercise regularly in Beihai and Jingshan parks in the morning where they can dance to music, an activity inappropriate in communal courtyards for fear of disturbing others.

There are two small, communal gardens in Suzhou Tongfangyuan (Case 4). One is located at the center for a kindergarten, and the other at its northwest corner, a place that seems not only to have been long neglected, but is also an unpleasant one to be in because of loud noise from the adjacent hotel's air-conditioners, along with distasteful cooking odors from nearby restaurants. The chosen tree type is also unsuitable as it blocks sunlight, with fallen leaves becoming wet and slippery. The property manager revealed that residents can be aggravated when talking about their communal garden because of these annoyances. Nevertheless, living adjacent to the famous Lion Grove Garden and in walking distance of the Humble

Figure 10.1   Beihai Park, previously an imperial garden, central Beijing. Photo by Yitong Lok 2010

Figure 10.2   Jingshan Park (Coal Hill) viewed from Beihai Park, Beijing. Photo by Yitong Lok 2010

Figure 10.3 Cultural Palace park just east of the Forbidden City, Beijing.
Photo by Donia Zhang 2007

Figure 10.4 Changpuhe Park ('Calamus River Park' or 'Iris River Park')
near Nanchizi new courtyard housing estate, Beijing. Photo by
Donia Zhang 2007

Figure 10.5   Lion Grove Garden adjacent to Tongfangyuan and Shilinyuan
               new courtyard-garden housing estates, Suzhou. Photo by Donia
               Zhang 2007

Figure 10.6   Little Flying Rainbow Bridge in the Humble Administrator's
               Garden near Tongfangyuan and Shilinyuan, Suzhou. Photo by
               Donia Zhang 2007

Administrator's Garden (both are listed as UNESCO'S World Cultural Heritage sites), Tongfangyuan residents can easily stroll in these city gardens.

A resident at Shilinyuan (Case 5) also noted that cultural activities, such as a private calligraphy club or morning exercises in a city garden, often take place outside the housing estate. As Shilinyuan is just east of Tongfangyuan (Case 4), it is also near the Lion Grove Garden and the Humble Administrator's Garden. However, the space of such a city garden is limited because Suzhou gardens are originally family gardens that are small in size. Now they have become public gardens which many people frequent, making them unsuitable for quiet walks.

Cultural Festivities Indoors and Outdoors

The survey of 290 residents in Beijing and Suzhou shows that they celebrate the following cultural festivals in a descending order of extent: (1) Spring Festival/ Lunar New Year (mean=3.3 on a scale of 1-5; n=168; average=1.91), (2) Mid-Autumn Festival (m=3.4; n=134; a=1.57), (3) Lantern Festival (m=2.94; n=103; a=1.04), (4) Dragon Boat Festival (m=2.52; n=87; a=0.76), (5) Winter Solstice Festival (m=2.24; n=87; a=0.67), (6) Double Ninth Festival (m=2.27; n=84; a=0.66), (7) Qing Ming Festival (m=1.83; n=81; a=0.51), and (8) Mid-Summer Spirit Festival (m=1.79; n=70; a=0.43). A significant number of residents also celebrate Solar New Year (m=2.58; n=92; a=0.82).

However, 70 interviews with residents (35 in each city) yield a slightly different order: (1) Spring Festival (94%), (2) Mid-Autumn Festival (59%), (3) Dragon Boat Festival (36%), (4) Winter Solstice Festival (29%), (5) Qing Ming Festival (21%), (6) Lantern Festival (20%), (7) Start of Summer (7%), and (8) Laba Festival (3%).

A regional variation in the festival celebrations was also observed. For example, Lantern Festival is more celebrated in Beijing (19%) than that in Suzhou (1%), whereas Winter Solstice and the Start of Summer are more celebrated in Suzhou (23% and 6%, respectively) than in Beijing (6% and 1%, respectively).

These findings suggest that only festivals signifying the turning point of a season such as Spring Festival, Dragon Boat Festival (originally related to Summer Solstice), Mid-Autumn Festival (around Autumn Equinox), and Winter Solstice Festival are well observed by residents, along with rituals related to ancestral worship such as Lantern Festival (or 'Spring Spirit Festival') and Qing Ming Festival (or 'Tomb-sweeping Day'). Certain festivals such as Blue Dragon Festival (1%) and Summer Lantern Festival (or 'Mid-Summer Spirit Festival') (1%) are no longer celebrated as much, possibly due to a lack of or shortage of private courtyard space to support these festivities.

Moreover, Western festivals such as Christmas, Solar New Year, Valentine's Day, Mother's Day (2nd Sunday of May), [1] and Father's Day (3rd Sunday of June), [2] are celebrated by contemporary China's younger generations (6%; n=70). Conversely, three of the 25 foreign residents in Beijing celebrate Spring Festival.

A wide continuity in Chinese cultural festival celebrations can be attributed to effective Chinese mass media such as radio and television. For example, CCTV has a rolling strip of information at the bottom of the television screen as a reminder of the arrival of a Chinese festival. Nevertheless, four of 70 interviewed residents commented that their small housing units often restrict important cultural festivities, such as a family reunion during Spring Festival.

However, the survey results show that only three *favorite* cultural festivities are celebrated in courtyards by the 290 survey respondents: Mid-Autumn Festival (1.4%), Spring Festival (0.7%), and Lantern Festival (0.3%). This finding implies that most residents conduct cultural festivities indoors, perhaps due to cold weather at Spring Festival and the lack of private courtyard for Mid-Autumn Festival.

The Pearson correlation results show a low, positive correlation between having a private yard and cultural *festivities* ($r=0.232$; $n=79$; $p<0.04$), but not cultural activities ($r=0.031$; $n=102$; $p<0.754$), although a high correlation ($r=0.676$; $n=172$; $p<0.000$) is observed between cultural festivities and cultural activities, suggesting the more cultural festivities are celebrated, the more cultural activities will likely be undertaken.

The Pearson correlation results also indicated no correlation between the number of households sharing a courtyard and cultural festivities ($r=0.049$; $n=72$; $p<0.683$) or cultural activities ($r=0.083$; $n=83$; $p<0.455$). Moreover, no correlation is observed between age and cultural festivities ($r=0.052$; $n=173$; $p<0.496$) or cultural activities ($r=0.046$; $n=211$; $p<0.506$) in courtyards.

Spring Festival Celebrations

Spring Festival (1st day of the 1st lunar month) celebrations remain more or less the same as before (Chapter 5) and was still enjoyed as the biggest festivity among the 70 interviewed residents, but certain details are simplified as modern elements become complex. For this festival, every household put on simple decorations such as pasting on door panels '福' (*fu*, meaning 'good fortune,' sometimes pasted upside down to resonate the sound of 'good fortune has arrived'), pinning traditional New Year scrolls (or posters, paintings, couplets) on interior walls, affixing window paper-cuttings and currency in fiery red colors (51%), hanging red lanterns in front

---

1   Since 2006, China has officially established its own Mother's Day on the 2nd day of the 4th lunar month, which was the day when Mengzi's mother gave birth to Mengzi and became China's most famous mother because of the education she offered to her son (Chapter 3).

2   Since 2006, China has officially recognized its own Father's Day on the 9th day of the 9th lunar month, overlapping with its Double Ninth Festival.

of homes (19%; some lanterns are modernized with light bulbs), displaying new calendars, placing fresh flowers in the dining room, and hooking up Chinese knots (3%). Some decorations may stay in place for up to a year (4%).

On Lunar New Year's Eve, some residents host a family reunion dinner at a restaurant (11%) and then sing at a karaoke (3%), while others celebrate at home (41%). With many dishes, the reunion dinner often includes meat, vegetables, wine, soft drinks, and *niangao* (or New Year pudding made of polished glutinous rice). Each household will leave all the lights on until midnight to expel 'evil.' Some families also worship their ancestors this evening (3%). After dinner, while watching CCTV's Spring Festival (Global) Gala, Beijing families start making dumplings with meat and vegetables (some with sterilized coins as filling), as well as other traditional New Year foods, such as steamed buns stuffed with sweetened red-bean paste with a red dot painted on top of each one signifying auspiciousness. On the other hand, Suzhou households make *tangyuan* (large stuffed dumplings made of glutinous rice flour served in sweet soup) and *hundun* (small dumplings in soup). At midnight, they will cook and eat dumplings, *tangyuan,* or *hundun* (23%). A few residents also mentioned they will also start to phone/call relatives and friends to relay New Year wishes.

At midnight, some residents let off firecrackers or fireworks in their courtyards (23%) or *hutong* if the courtyard is too small (Case 1), give red envelopes with money to the children (3%), and stay up all night playing *majiang* (4%). Since several new courtyards at Nanchizi (Case 2) have wooden vine trellises that dry up in the winter, firecrackers will not be let off in it for fear of fire hazards. Nonetheless, a resident noted that more households at Juer Hutong (Case 3) let off firecrackers on their balconies in the most recent 2 years (2009-2010) than in the past, whereas a resident at Shilinyuan (Case 5) observed a significant decrease in these festivities in recent years because people seek quietness and avoid disturbance.

On Lunar New Year's Day and the days after, some residents visit their relatives (9%) while others go to a temple market (*miaohui*, or 'festive market') (3%), where they buy traditional foods and drinks for entertainment. Still others go to a Buddhist temple and pray with lit incense for a lucky year (1%), or travel to a new place (1%). These festivities last until the 5th day of the Lunar New Year with ancestors' photos often displayed at home until the 15th day (1%).

A resident at Jiaanbieyuan (Case 6) observed that since the quality of life is improved, everyday is like a festival, everything is convenient to buy, and there are only things that one cannot think of, but nothing one cannot buy. Thus, the improved quality of life may contribute to a lesser extent of New Year celebrations than that in the past.

Three of 25 surveyed foreign residents in Beijing celebrated Spring Festival by pasting '福' and hanging red ribbons or auspicious banners on their door panels. A Chinese resident at Juer Hutong (Case 3) recounted that once during Spring Festival, some foreign residents served a feast in the courtyard in very cold weather wearing thin clothes. They seemed to endure the cold well and like to join the festive atmosphere of the city, an obvious effort in trying to achieve social cohesion.

Figure 10.7   Red lanterns hung from the shared stairwell of a communal courtyard at Juer Hutong new courtyard housing, Beijing. Photo by Donia Zhang 2007

Figure 10.8   Apartment doors decorated with 福 ('good fortune') or its inverted form, door gods, or paired lucky dolls that may have stayed since last Spring Festival, Juer Hutong new courtyard housing, Beijing. Photo collage by Donia Zhang 2007

Figure 10.9 A spring couplet hung on the doorway of an apartment at Juer Hutong new courtyard housing, Beijing. It reads: "May everything you wish come true with the arrival of spring; may good fortune enter this house with an abundant harvest of crops" (万事如意新春始，五谷丰登福即来), and the horizontal strip says: "Welcome good fortune and greet auspiciousness" (纳福迎祥). Photo by Donia Zhang 2007

Spiritual/Ancestral-Worship Festival Celebrations

Traditional ways of celebrating Lantern Festival and Qing Ming Festival are preserved to a large extent, but by only a small number of residents.

On Lantern Festival (or 'Spring Spirit Festival,' 15th day of the 1st lunar month), residents eat *yuanxiao* (glutinous rice balls filled with sweet fillings) at home or in a restaurant (14%; n=70). Some families then go to a temple market or a city park to view lantern exhibitions (3%). Several residents reported that they do not hang lanterns at home anymore although they did so in childhood (3%).

On Qing Ming Festival (April 5, or April 4 in leap years), residents groom their ancestral tombs and offer sacrifices to deceased elders (14%). Some also climb hills and fly kites to send away worries (1%).

Summer Festival Celebrations

On the Start of Summer (May 5-7, one of the 24 solar divisions), Suzhou residents eat salted duck eggs, hang wormwood on their door panels (9%; n=35), and don a salted duck egg in a string bag around a child's neck to prevent parasites from harming the child's health (3%).

On Dragon Boat Festival (5th day of the 5th lunar month, originated from Summer Solstice, June 20-21), families gather to eat *zongzi* (a pyramid-shaped dumpling made of glutinous rice wrapped in bamboo or reed leaves) at home (30%; n=70). Some Suzhouers wrap up wormwood, calamus, and garlic in red paper and hang the roll on their door panels to prevent insects from harming their health (6%).

Mid-Autumn Festival Celebrations

Mid-Autumn Festival (or 'Moon Festival,' 15th day of the 8th lunar month) celebrations are sustained to some extent, with modern elements added. On this festival, families reunite at home or in a restaurant to celebrate harvest and eat seasonal fruits and moon cakes (44%; n=70) while watching CCTV's Mid-Autumn Festival (Global) Gala (7%). Some Suzhou residents also share water chestnuts, sweetened taro, and a rich variety of foods (3%).

When Beijing's traditional courtyards were big enough to hold a table and chairs, residents served plates of fruits and moon cakes from the table while enjoying the full moon. Now there is no space in the courtyards for such festivities. Residents expressed that they would love to look at the moon in a proper courtyard because it would put them in the right mood for it (22%; Case 1).

Residents at Nanchizi new courtyard housing (Case 2) are fortunate to have relatively large courtyards, nevertheless, only a small number of them appreciate the full moon in the courtyards (17%). A resident described that at around 6 pm on the night of Mid-Autumn Festival, neighbors will sit in the courtyard chatting and savoring grapes, moon cakes, and roasted meat while enjoying the full moon together. In a private courtyard at Nanchizi, all in-laws will come together with sometimes more than 10 people from three generations to barbeque and share foods in the courtyard. They feel more comfortable using a private courtyard than a communal one because neighbors will not complain, and there are fewer constraints for festivities.

Some residents at Juer Hutong new courtyard housing (Case 3) spend Mid-Autumn Festival at home watching the full moon from their windows (21%), while others go boating in Beihai or Shichahai Lake District (21%). Hardly anyone will be in the courtyard that evening except several children at play. This confirms my observation on the Mid-Autumn Festival (September 25) in 2007 when two boys played in Courtyard A from 6-7 pm, but *no one* sat in any of the courtyards. An elderly resident noted that the new courtyards are unlike traditional single-extended-family courtyards where every family member would come out to enjoy the full moon. Another elderly resident revealed that he had just moved in 2 weeks previously with his family and found that except for his two grandchildren who sometimes play in it, the communal courtyard is so strangely quiet during the day and even on the night of Mid-Autumn Festival. Thus, private ownership of a courtyard may enhance its usage for cultural festivities.

In Suzhou, some residents admire the full moon through a window at home if the weather is fine (6%). If a household has a private courtyard, they will place a table in it (9%), with five plates offering seeds, candies, and lotus to the moon goddess as lotus symbolizes the moon's purity. If there is no private courtyard, a family will celebrate with the same offerings on the balcony or in front of their gate. Some residents may take a romantic 'moonlight walk' (*zouyue*) along the rivers' long corridor encircling the city, or row a boat where they can view the full moon without visual interference (9%). Hence, there are regional variations in Beijing and Suzhou on how Mid-Autumn Festival is celebrated. Nevertheless, an outdoor setting is essential to fully experience and to appreciate such a festival.

Winter Solstice Festival Celebrations

Although Beijing and Suzhou residents still celebrate Winter Solstice (December 21-22), they mostly do so indoors due to cold weather. On Winter Solstice, Beijing residents usually make and eat dumplings (9%; n=70), while Suzhou respondents typically drink winter wine (9%). In Suzhou, Winter Solstice is almost as important as Spring Festival (9%) with a local proverb saying, "Winter Solstice is bigger than Spring Festival; the rich eat all night while the poor suffer cold."

Winter Solstice is also a time for family reunion (11%). On the night before, Suzhou families prepare a table of many dishes with osmanthus winter wine (a traditional Suzhou rice wine made of osmanthus), New Year pudding or lamb pudding (made of frozen lamb mixed with lamb soup to keep one warm), and vegetables (17%).

As well, Winter Solstice is a time for remembering ancestors with a ceremony (11%) of lighting fragrant candles and burning tinfoil paper as funeral offerings to pay respect to deceased family members before the living descendants can sit down together for dinner (14%). Nevertheless, some families now prefer a restaurant meal (3%) as festivities are modernized and simplified.

Birthday Celebrations Indoors and Outdoors

Holding a birthday party is the 5th major cultural activity in courtyards, but only 3 percent of 290 survey respondents considered it their *favorite* cultural activity in courtyards. Still, 47 percent of 70 interviewed residents celebrate their birthdays, typically featuring a special meal at home (31%) or in a restaurant (17%), and perhaps also singing at a karaoke (3%). The celebratory meal often includes two boiled eggs for the year to go smoothly (14%), noodles for a long life (13%), and a birthday cake (13%) in the shape of a peach to symbolize longevity (3%). Thus, traditional birthday celebrations are more or less maintained with added modern elements.

One resident revealed that he celebrates his birthday by inviting his mother for a meal because of the proverb: "A child's birthday joy is the mother's toil." Senior family members' birthdays receive more attention, especially the 50th, 60th, 70th, 80th, and 90th year mark (10%). Others celebrate only every 5th and 10th birthday (3%). If a family member's lunar birthday happens to be on a festive day, they tend to combine the celebrations together (3%). Some families are specific about the festive procedure that the day before the birthday, they will eat dumplings to send away the old year, and eat noodles and a cake on the birthday for longevity (1%). Some grandparents buy new clothes for a grandchild's birthday (1%) and conversely, some children choose the same for a parent's birthday (1%). In one courtyard compound at Nanchizi new courtyard housing (Case 2), residents sing "Happy Birthday to You" to neighbor to reveal a degree of social cohesion created by the communal courtyard on this occasion.

Wedding Ceremony/Anniversary Celebrations

The survey of 290 residents in Beijing and Suzhou show a low mean and average value (m=2.02 on a scale of 1-5; n=61; a=0.42) of people holding a wedding ceremony in the courtyards, and none of the residents considered this event as their *favorite* cultural activity. The fact that a wedding is a rare activity may well be why people did not rank it as popular. As well, few residents may actually get married during the time they live in these courtyard housing estates because the average age of survey respondents was 50 in 2007, and as the new courtyard housing estates were built in 1990-2003, many would have got married before moving into these estates.

A resident, who had lived at Juer Hutong new courtyard housing (Case 3) since its inception, noted that no one holds a wedding ceremony in the communal courtyards. If a marriage occurred in the housing estate, the newlyweds would send wedding candies to neighbors after the ceremony. Thus, a traditional ceremony in the courtyard is almost a lost ritual, possibly because the courtyard is too small, and in most cases, not a private space anymore for intimate events as such.

Nonetheless, 27 percent of 70 interviewed residents celebrate their wedding anniversaries at varying degrees, with celebrations often including a romantic meal somewhere (16%), gift exchange at home (4%), or a scenic tour (3%). Some residents celebrate only every 5th and/or 10th wedding anniversary (3%) such as 25th (3%), 30th (4%), 50th (9%), and 60th (6%) year mark. Beijing and Suzhou residents adopt Western notions of 'silver (25th),' 'gold (50th),' and 'diamond (60th)' for wedding anniversaries celebrations.

Summary and Conclusion

This chapter examined residents' cultural beliefs and behavioral patterns in the renewed/new courtyard housing in Beijing and Suzhou, and explored which factors would help or hinder their use of the courtyard. The findings suggest that Communism has managed to eliminate most residents' traditional faiths, although Chinese philosophy is still deeply ingrained in the minds of a small number of residents.

Modern lifestyles also largely affect residents' behavioral patterns. Although indoor activities seem to take up most of a resident's time each day, the communal courtyards help sustain some traditional Chinese cultural activities, but not as much as the single-family private courtyards. The primary function of a communal courtyard is to maintain health/natural healing; it facilitates harmony with oneself, or self-cultivation. However, many cultural activities are much less or no longer partaken in the communal courtyards, likely due to such factors as time, climate, courtyard ownership, yard size, facilities, and so on. Moreover, community/city parks/gardens have become important places for cultural activities in China.

Stepanchuk and Wong (1991) argue that one of the best ways a nation may sustain its cultural identity alongside the homogenizing effects of globalization is through continued celebrations of its cultural festivals (p. 109). However, only major traditional Chinese cultural festivities related to seasonal change are still performed. Other less significant traditional festivities are only conducted by a small number of residents. In addition, festivals connected to ancestral worship are also celebrated by only a small number of residents. While these festivities have combined traditional and modern characteristics and are mostly celebrated indoors, a courtyard is still instrumental to viewing the full moon on Mid-Autumn Festival, and letting off firecrackers/fireworks on Spring Festival (with some debates over this fiery tradition). The reduced degree of cultural festivities and wedding ceremony in communal courtyards may be due to modern lifestyles, and a lack of private courtyard spaces for intimate family events.

In Chapter 11 that follows, a synthesis of the empirical findings and literature review will be presented, in an attempt to answer the main research questions.

PART IV
Culturally Sustainable Architecture

Chapter 11

Four Cornerstones of Culturally Sustainable Architecture

In the beginning was the Dao.
All things issue from it;
all things return to it.
To find the origin,
trace back the manifestations.
When you recognize the children
and find the mother,
you will be free of sorrow.

– Laozi, c.571-471 BCE, *Dao De Jing*, Verse 52

This chapter synthesizes the findings and attempts to answer the main research questions. It takes the stance that the four key themes identified in Chinese philosophy: harmony with heaven, harmony with earth, harmony with humans, and harmony with self (Chapter 3), are benchmarks against which the cultural sustainability of the renewed/new courtyard housing in Beijing and Suzhou can be measured (Chapters 7-10). This information allows me to theorize the concept of culturally sustainable architecture in China, which is the focus of the chapter.

Sustainable architecture has been a popular subject of study promoted by many professionals since the 1990s. It also has many different names, for example, 'sustainable design/development' (Bergman, 2011; Earth Pledge, 2001; Lechner, 2008; Lyle, 1994), 'green design/development' (Attmann, 2009; Carroon and Moe, 2010; Earth Pledge, 2001; Gissen, 2003; Johnston and Gibson, 2008; Kibert, 2011; Kwok and Grondzik, 2007; Stang and Hawthorne, 2010; Wagner, 2008), 'regenerative design' (Lyle, 1994), 'biodesign' (Moore, 1995), 'ecological/environmental design' (Hester, 2006; Orr, 2002; Yeang, 1995), and 'eco-house/eco-housing' (Moro and Spirandelli, 2011; Roaf, Fuentes, and Thomas, 2007; Wagner, 2008).

A common theme of these publications is the focus on 'green' issues and the technological aspects of architecture regarding efficient energy consumption, earth-friendly materials, and sustainable construction. Their emphasis is on maximizing daylight, solar heating, passive cooling, and minimizing waste by using recycled materials and water conservation features. Sustainable architecture is also about creating a healthy and comfortable environment by reducing operation and maintenance costs, and addressing such issues as historic conservation, access to public transportation, and other community infrastructures. The entire lifecycle

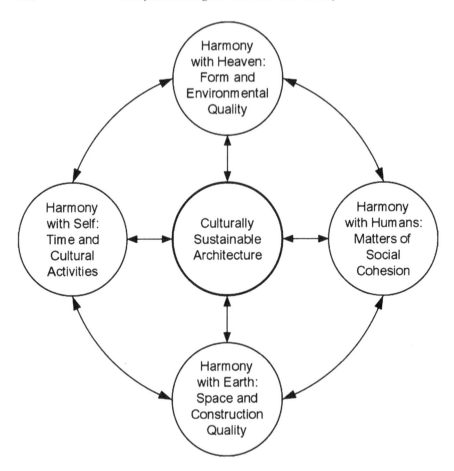

Figure 11.1   A conceptual model of culturally sustainable architecture in China. Drawing by Donia Zhang 2011-2012

of the buildings and their components are considered, as well as the economic and environmental impacts and performances (East St. Louis Action Research Project, n.d.; Kwok and Grondzik, 2007; Lyle, 1994; Paehlke, 2004; Sassi, 2006).

Some authors have raised social and cultural issues relating to sustainable architecture and argue that sustainable architecture should create a balanced life for its occupants within nature and culture, and that it represents critical, connected pieces of the whole essential to human health, social, cultural, ecological, environmental systems, as well as regional economy, while respecting the laws of nature (Earth Pledge, 2001; Hester, 2006; Liang, 1998; Schmid, 1987; Wheelwright, 2000; Williams, 2007). Sustainable architectural forms may foster changes in human behavior that potentially enable citizens/neighbors to connect and get to know each other in their localities, and where community members can work together effectively to solve complex problems. Sustainable architecture can also heighten an awareness of interdependence and concern for others in the community (Hester, 2006, pp. 16-19). However, sustainable architecture may only be possible with rigorous scrutiny of our cultural histories, environmental theories, and technological practices because traditional architecture provides a rich source of inspiration for us (Khattab, 2002; Liang, 1998; Pallasmaa, 2001; Wheelwright, 2000; Wu, 1999; Yu, 2007). Based on traditional Chinese philosophy, this research suggests that culturally sustainable architecture rests on four cornerstones:

1. Harmony with heaven: the form and environmental quality that offer maximum climatic protections, good sunlight, air, and view of nature;
2. Harmony with earth: the space and construction quality that withstand the test of time without requiring short-term repairs or demolition;
3. Harmony with humans: matters of social cohesion that demand nurture from the built environment for good social relations; and
4. Harmony with self: time and cultural activities that necessitate the built environment to facilitate self-cultivation.

The inclusion of human dimensions establishes a systematic framework to measure culturally sustainable architecture in China.

Harmony with Heaven: Form and Environmental Quality

A classical Chinese courtyard house was a link between heaven, earth, and humans. *The Yellow Emperor's Canon on Houses* (黄帝宅经) states: "A house is the hub of *Yin* and *Yang* and the image of the universe…a person becomes independent because of his/her house, and a house survives because of its owner, as human and house sustain each other, Heaven and Earth communicate with one another"[1] (Ma, 1999, p. 30; Luó, 2006, p. 10; Luò, 2006, pp. 1, 104; my translation). *The Book of Three Principles* (三元经) also indicates, "If the house is auspicious, its inhabitants will be prosperous"[2] (Luò, 2006, pp. 1, 104; my translation).

Exterior Form

For a house to be auspicious, traditional Chinese philosophy considers that its exterior form is of primary importance because it often determines the environmental quality (light, air, sound, view, etc.) that ultimately affects the occupants' health. There should be abundant sunlight on site, but not too strong wind or it would cause illness. The site should be far from sources of pollution, including in the air (dirt, fumes, grime, haze, smog), water (contaminants, toxins),

---

1   夫宅者,乃是阴阳之枢纽,天伦之轨模…人因宅而立,宅因人得存,人宅相扶,感通天地。
2   地善即苗茂,宅吉则人荣。

light (excessive infrared and ultraviolet radiations), noise (automobiles, factories), as well as electromagnetic (radars), thermal (boilers), and visual types (wastes, advertisements), and so on that may impinge on the occupants' welfare (Kou, 2005, pp. 81, 165; Kwok and Grondzik, 2007, p. 141; Luò, 2006, pp. 124-133).

The World Health Organization estimates that air pollution is associated with 600,000 annual deaths globally (Guo, 2006). According to the United Nations, respiratory disease is China's top killer, especially among children (Guo, 2006). Guo (2006, p. 27) argues that while industrialization shares part of the blame, motorization affects urban residents most directly with massive increases in paved areas causing severe rainwater runoff, creating another major toxic source. Hence, greening the cities with more courtyard-gardens may be one way to solve the problem.

It has been argued that the environmental crisis is mainly a result of humans having disconnected with nature by living in artificial environments so extensively that we may have lost our sense and sensibility needed to observe symptoms when natural processes become dysfunctional (Lyle, 1994, p. 100; Orr, 2002; Zhang, 2009). Culturally sustainable architecture allows open windows to let in air and sunlight, uses solar heating in interior environments, and rejoins human shelters with nature's roles (Lyle, 1994, pp. 40, 100). In this sense, architectural forms have less to do with style and more to do with the site, light, and wind (Lyle, 1994, p. 109).

This study likewise shows that the top five reasons for 290 surveyed residents in Beijing and Suzhou to buy/rent a new courtyard housing unit are environmental (39%), cultural (28%), relocation (25%), aesthetic (24%), and location (17%). Thus, environment is the most significant factor for residents in deciding where to live, and good environmental quality seems to be associated with the courtyard form (Chapter 2).

A classical Chinese courtyard house was not only a sanctuary in which to worship Heaven, but also a home to accommodate an entire family clan (Chapters 3-5). However, contemporary economic, social, and cultural factors no longer support such living patterns. The family structural change from single-extended to 2-3 persons nuclear family, and the emergence of varied household types such as single-parent, DINK ('double income and no kids'), unmarried, and so on, have demanded smaller housing unit designs (Gao Zhi, interview, 2008). To sustain traditional courtyard housing forms and appropriately modernize interior circulation systems is an important but challenging task facing Chinese planners and architects today. There is a view that it is unnecessary to painstakingly preserve traditional Chinese dwelling culture because modern people have unconventional habits and lifestyles. Instead, architectural designs should respond to current changes by following demands that result from new urban lifestyles (Gao Zhi, interview, 2008; Liu Li, interview, 2008). This study shows that while some residents value the aesthetics of traditional-style exteriors, most are indifferent to them. This finding confirms that of Fang (2004, 2006).

Exterior Walls

Classical Chinese courtyard houses had enclosing walls surrounding the entire compound; they were of grey or red colors in Beijing and white in Suzhou (Chapter 4). Materials for the walls were solid and durable to reduce maintenance costs in the long run. Lattice or assorted window patterns were replicated on the enclosing walls in Suzhou to invite the admiration of passersby of courtyard-gardens (Kou, 2005, pp. 54, 64; Luò, 2006, p. 139). The enclosing walls continue to be a feature valued by Beijing residents (Zhao, 2000).

This research shows that residents appreciate the exterior walls (building envelope) in traditional materials and colors, but some walls do not always perform well because of the new construction methods employed. Exterior walls demand good insulation to reduce energy consumption, and good-quality construction to offer protection from all weather conditions such as rain, storm, and snow (Kwok and Grondzik, 2007, pp. 23, 25).

Gate and Access

In traditional Chinese philosophy, a gate as the 'mouth of *qi*' in a southeast orientation is critical to the occupants' health (Chapter 3). This study shows that a significant number of residents in Beijing and Suzhou still value a traditional southeast/south (51%; n=290) orientation for their gate because they believe that 'purple *qi* comes from the east,' or the sunrise direction (Chapter 7).

For safety and convenience, *Feng Shui* theory suggests that a house compound should have two gates, and that they should not exactly line up with each other, in order to reduce the likelihood of strong cross winds that may cause diseases, particularly for women (Kou, 2005, pp. 91, 94; Luò, 2006, p. 172). Suzhou Tongfangyuan new courtyard-garden housing estate complies with this theory by having the south gate slightly offset the north gate, with the residents satisfied with one gate in each of the four cardinal directions (east, south, west, and north). For larger housing estates such as Jiaanbieyuan, two gates in the south and north respectively are inconvenient to its residents who prefer easier access via a gate in each of the four directions.

Windows

In traditional Chinese philosophy, sunlight is an important factor in window design because of its many advantages: bringing warmth to people, enhancing vitamin D's synthesis in the human body, preventing children from developing rickets, and slowing the speed of osteoporosis in elderly people, not to mention that ultraviolet rays in sunlight can kill bacteria, and can strengthen the human immune system. Nevertheless, too much sunlight may also cause health ailments, including skin cancer (Kou, 2005, pp. 21, 169; Luò, 2006, pp. 163-164).

Feng Shui theory thus advocates that buildings should ideally be 'sitting north and facing south' for more sunlight and better natural ventilation (Chapter 3). Modern concepts of green design likewise suggest placing window openings in the north-south directions to gain more solar heating and day lighting (Kwok and Grondzik, 2007, p. 23; Luò, 2006, p. 161; Lyle, 1994, p. 112).

As well-controlled distribution of sunlight is a basis of green design, skylights or top lighting will offer uniform distribution of natural light across large areas of an interior space, and will contribute to energy performance, residents' health, and their contentment. Besides, the amount of solar rays admitted to a building needs to be adjusted throughout a year via movable shading devices that can provide flexible and responsive control to heat gains, particularly for west- and south-facing windows (Kwok and Grondzik, 2007, pp. 23, 55, 139-140; Luò, 2006, pp. 161-162; Lyle, 1994, pp. 108, 112-115). This research indicates that French windows in Beijing Nanchizi new courtyard housing present problems for residents because they lose too much heat in the winter but gain too much of it in the summer. Thus, French windows are climatically unsuitable for China's northern regions.

Under appropriate climatic conditions and potential differences in the indoor-outdoor air temperatures, cross ventilation is a practical, energy-efficient alternative to mechanical cooling. Functional cross-ventilation demands a building form to maximize exposure to the prevailing wind directions (Kwok and Grondzik, 2007, pp. 23, 55, 139-140; Luò, 2006, pp. 161-162; Lyle, 1994, pp. 108, 112-115), which requires north-south or east-west window orientations in China (Kou, 2005, p. 97).

This study shows a significant number of Beijing and Suzhou residents still value traditional north-south cross-ventilation (78%; n=290). Some residents also prefer a window in each of the four cardinal directions to gain maximum sunlight, ventilation, and views of sunrise and sunset. If there is only one window, they still prefer a south-facing one (74%; n=290). These preferences relate to a perceived better environmental quality, even though only 40 percent of 290 survey respondents have south-facing windows. Subsequently, only a few interviewed households have mentioned that they have windows in all four directions.

Courtyards and Gardens

Traditional Chinese houses generally had courtyards and sometimes also gardens (Chapter 4) that offered the occupants direct contact with the sky and green trees of native species that helped release stress from life (Kou, 2005, pp. 30-38, 49; Luò, 2006, pp. 105, 146). *Feng Shui* theory suggests that a courtyard size should be proportional to the height of surrounding buildings to admit sufficient sunlight (Chapter 3).

China's *Design Code for Residential Buildings* (1999, 2003) also stipulates that every unit should have at least one room that receives a minimum of 2 hours of sunlight on the Great Cold Day (January 20-21) (Luò, 2006, p. 134). However, this research shows that many housing units in the case study estates have not met this criterion due to small courtyard sizes. For new courtyards to receive an equal amount of natural light as in traditional courtyards, the ratio of building height to distance is minimum 1:3 for Beijing (Zhang, 1994, 2006, 2009/2010/2011) and 1:1.3 for Suzhou (Chapter 7).

The research shows that Chinese city planning policies have endorsed the continuity of courtyard form but restricted architectural prototypes with detailed regulations to control the density and plot ratio, leading to negative consequences on courtyard size designs. Besides, corruption is frequently cited as a cause of adjusted and smaller courtyards, which goes against Confucian ethics which suggests that personal gains should not be obtained at the cost of morality or justice (Gong, 1991, p. 318; Kuang, 1991; p. 12).

Roofs

Traditional courtyard houses in northern and eastern China normally had pitched roofs to protect the occupants from heat, cold, rain, wind, and snow (Chapter 4). Pitched roofs are also more visually appealing than flat ones (Lyle, 1994, p. 115; Rapoport, 1969, p. 134). Flat roofs have a major problem in maintaining occupants' thermal comfort (Lyle, 1994, p. 116). The findings of this research indicate that residents prefer a pitched roof over a flat one because of its better thermal performance rather than its cultural or aesthetic appeal.

Discussion

These case projects are not always in harmony with heaven because they cannot always provide residents with maximum physical protection or comfort under any weather condition, although the Suzhou cases generally perform better than the Beijing ones. The symmetrical planning along a central axis linking Heaven and Earth in Chinese philosophy has not been observed in the designs of Nanchizi, Juer Hutong, or Jiaanbieyuan housing estate. Conversely, only a small proportion of residents expressed concern for exterior appearance of the housing estate, while function was seen as fundamental by most residents.

Harmony with Earth: Space and Construction Quality

Humans spend at least one third (8/24 hours per day) of their life at home (Kou, 2005, pp. 6, 110; Luò, 2006, pp. 1, 193), and most people spend a minimum of 90 percent of their time indoors (Lyle, 1994, p. 112). Home is where one would feel at rest, with a Chinese saying expressing it well: "Neither a gold nest nor silver nest is as good as my nest." The concept is similar to the English phrase, "East, west, home is best." Hence, interior space is of paramount importance.

Interior Space

Traditional Chinese philosophy suggests that the size of a house should be proportional to the size of a household (Chapter 3). In his book, *Feng Shui and Modern Residences* (2006), Luò indicates that for a 3-person unit of 150-180 sqm, the living or dining room should be 25-40 sqm because the living room is the hub for family activities and the guests, and the dining room is the focal space where family and friends gather to eat and communicate, especially during special occasions. As bedrooms are the most important functional spaces in a home, a master bedroom of 15-25 sqm is adequate. Since a reading room is where one studies, works, thinks, and makes important decisions, it needs plenty of light and oxygen. A kitchen of 6-15 sqm and a bathroom of approximately 10 sqm are large enough for their intended functions and basic facilities. Corridors, staircases, and hallways should be no less than 90 cm for occupants' movement and shifting of furniture (Kou, 2005, pp. 114, 122; Luò, 2006, pp. 107-108, 150-153, 157-158, 174).

This research shows that although many units at Suzhou Shilinyuan and Jiaanbieyuan have met the above design criteria, most units in the Beijing cases have not. This discrepancy may be due to their planning policies and more rigorous construction requirements set by the Suzhou municipal government. As Beijing's renewed traditional courtyard houses have not solved the problem of small living spaces, residents frequently build extra rooms in communal courtyards that further block sunlight and ventilation, while creating fire hazards and blocked escapes in case of earthquakes. Unlike classical Chinese courtyard houses (Chapter 4), the interior spaces in the new courtyard housing at Beijing Nanchizi and Juer Hutong are so restrictive that many residents complained about impractical and irrational designs, particularly the small unit sizes of 45-60 sqm for 2-storey housing at Nanchizi with steep, dangerous staircases. Several residents at Juer Hutong, Tongfangyuan, and Jiaanbieyuan also commented that their staircases in the middle of the hallway in duplex apartments, and a column inside a room make it awkward to place furniture. Nevertheless, the elimination of socio-spatial hierarchy in these new courtyard housing designs is seen to represent China's social progress.

The floor to ceiling height is best to be 2.8-3 m for healthy air circulation, while the common 2.5-2.8 m ceiling height in China is low (Luò, 2006, p. 169). Residents in this research indeed criticized their low ceilings below 2.8 m (Chapter 8).

Feng Shui theory suggests that it is inappropriate to place a kitchen or bathroom at the center of a housing unit because damp or stale air from them may stagnate indoor air and harm the occupants' health. The door to the kitchen or bathroom should not directly face an opening to living room, dining room, or bedroom for the same reason. The stove should not directly face the kitchen door so that wind may not extinguish the cooking fire. A kitchen and bathroom must also have a window leading to the outside for fresh air (Kou, 2005, pp. 77, 132, 134, 140; Luò, 2006, pp. 135, 168, 181-183, 188). However, some kitchens and bathrooms in the estates studied do not have windows for natural ventilation.

Floor Levels

Traditional Chinese philosophy advises that people should live close to earth *qi* for better health, but not digging/gouging ground to hurt Mother Earth (Chapter 3). Therefore, classical Chinese courtyard houses were mostly designed with 1-3 storeys without basements (Chapter 4). Ample empirical studies also indicate that housing should not have too many storeys for occupants' safety and wellbeing (Chapter 1). This research shows that Beijing and Suzhou residents predominantly prefer to live in housing of 1-3 storeys for practical reasons. This finding supports that of Wu (1999) and Zhao (2000). However, residents' present low-rise living has a low correlation with their preferences, suggesting that habit may affect floor level preference.

Furniture Styles and Materials

Classical Chinese interior furniture was always made of natural material such as red/hard wood that has a warm touch (Chapter 4). This research shows that a large number (40%; n=67) of Beijing and Suzhou residents still value traditional Chinese-style red/hard wood furniture for their homes. Some explained that they like their furniture style to match that of architecture for consistency of appearance. Nevertheless, modern design aphorisms of 'less is more' and 'beauty is in simplicity' were also appreciated by a significant number of residents (27%) who chose modern Western-style furniture. Contrary to findings of Kai-Yin Lo (2005), this research did not find a notable *trend* of following Western décor among residents in either Beijing or Suzhou.

Facility Provision

Classical Chinese courtyard houses were not equipped with comfortable facilities by modern/Western standards (Chapter 4), partly due to technological developments at an earlier time, and partly due to Confucian ethics that advocate frugality and simplicity in life, such that humans should focus on their responsibilities rather than indulge themselves with material goods.

However, modern lifestyles have changed this traditional Chinese cultural value. This research indicates that when contemporary Chinese people demand all the basic service facilities at home (running water, electricity, gas, heating, air-conditioner, cables, wires, etc.), at the time of the survey (2007), some of Beijing's renewed traditional courtyard houses still relied on coal burning for winter heating, which is not only energy-inefficient but also pollution-generating and involves a risk of carbon monoxide poisoning. Moreover, many households did not have a private bathroom or kitchen for personal convenience. This finding confirms those of Lu and He (2004) and Wu (1999).

The research also shows that a bathroom is typically absent on the 2nd floor of housing units at Nanchizi and duplex apartments at Juer Hutong, making it awkward to use it at night. Moreover, drainage pipes across the cases studied are often so narrow that they cause frequent blockages and flooding inside the apartments.

Building Materials and Construction Quality

Traditionally in China, building materials and construction quality were regarded of utmost importance, as seen in the fine workmanship and elaborate details of classical Chinese courtyard houses (Chapter 4). Because buildings consume a huge amount of raw material resources (Ahmad and Zong, 1995; Lyle, 1994), wise application of them may have a major impact on the natural environment. Local, natural resources such as wood generally require less processing and are less energy-intensive than manufactured materials, especially metals (Lyle, 1994, p. 107). Nevertheless, different materials have different capacities for storing heat. For example, brick, stone, and water make outstanding storage for thermal mass, whereas wood, wallboard, and most metals have less heat-preserving properties (Lyle, 1994). As large glazing areas (French windows, solaria, and sunspaces, etc.) require large areas of thermal mass, the ratio of thermal-mass areas to glass surface should normally be minimum 3:1 (Lyle, 1994, p. 107), with insulation being a critical part of green design to reduce energy consumption (Kwok and Grondzik, 2007, p. 23). However, this research shows that Nanchizi new courtyard housing is poorly insulated, and that French windows are performing badly.

Building materials should be selected not only for their adequacy in construction, but also a concern with time dimensions and weathering. The primary matter is not what the building's style is, but how it responds to the material conditions surrounding it (Rapoport, 1969, pp. 113-114; Theis, 2005, p. 109). A truly sustainable construction will use long-lasting materials and reuse them repeatedly. Materials such as bricks, stones, and heavy timbers are often reused, whereas glass and broken concrete can be reprocessed but are rarely recycled (Kwok and Grondzik, 2007, p. 23; Lyle, 1994, p. 124).

Construction quality assurance should be of prime importance because no matter how fine a building's design is, any construction project will ultimately be assessed on the final product's competence. Good-quality construction could significantly reduce building costs and protect homeowners' investments (Fryer, 2007, p. 5; Harrison, 2005, p. 2). Construction quality is recognized as a critical issue in China (Ahmad and Zong, 1995; China Daily, 2010-04-12; Travel China and the World, 2010-11-01; Yung and Yip, 2010). Reasons for low-quality constructions can be attributed to short-term planning of a 20- to 30-year building lifespan by the Ministry of Housing (China Daily, 2010-08-07), poor designs, inferior materials, weak construction management, and a zealous completion timeline, as well as developers' drive for profit, workers' skill inadequacies, and so on (Ahmad and Zong, 1995). My field survey reveals that contractors for the Beijing cases were mainly untrained, transient workers from rural China who lacked the necessary skills to undertake serious construction work, and that they may receive lower wages than regular city workers.

At the three Beijing cases, there is an obvious lack of construction quality assurance, and the poor-quality construction overshadows and damages the reputation of these new courtyard housing experiments. This finding confirms that of Liang and Zong (2005) on the Juer Hutong project. The construction quality problems are Beijing residents' major concern rather than the enjoyment and use of the courtyards. The construction quality is much better at the three Suzhou cases (Chapter 8), which may be due to a stricter construction quality requirement enforced by the Suzhou municipal government.

Maintenance and Management

This research shows that maintenance and management constitute a serious problem in the two Beijing cases due to inadequate professional training of public housing management personnel at Nanchizi, and to the difficulty of collecting maintenance fees from the residents at Juer Hutong.

The Juer Hutong new courtyard housing was built in the early 1990s when China still had a socialist system. As there was no such practice as property management, no maintenance fees were collected. Currently, residents have not entirely shifted their thinking from a planned economy to a market-driven one in which they have to pay for housing services, not to mention that some households may not be able to afford them. With no maintenance fees paid, the quality of management is poor, and because the quality of management is poor, residents are reluctant to pay any fees, keeping the system in a vicious circle. Although this finding conforms to that of Sheng (2008), the problem is less acute in the Suzhou cases because they employ private housing management companies. Nevertheless, Suzhou residents still complained that their private management companies are unwilling to seek property repairs when a fee is involved (Chapter 8).

Car Park Spaces

Classical Chinese courtyard houses normally had a carriage (or sedan chair) room in the southwest corner of the south hall adjacent to the *hutong* (Chapter 4). However, modern lifestyles have changed this situation.

An outcome of China's rapid economic development is the uncontrolled growth in private car ownership in cities since the 1990s (Peng Hongnian, interview, 2008). Statistics show that China has become the world's largest auto market: it surpassed Japan in 2006, and exceeded the United States in 2009 (Economist, 2009-10-23; RIA Novosti, 2011). Thus major Chinese cities such as Beijing have been given a nickname '*ducheng*' ('congested city'), portraying modern aspirations and with them, challenges to courtyard living.

Across all six cases in Beijing and Suzhou, there is a severe shortage of car park spaces because there were fewer private cars in China at the time they were built. This scarcity may also be linked to land use, as on each plot of land, non-built spaces (such as courtyards) in the middle of the built-up areas count as space and consequently, cars have much less space for parking in these designs.

Discussion

These case projects are not always in harmony with earth due to inferior materials and poor-quality construction that lead to countless post-occupancy repairs and post-repair waste disposals. China's Construction Law became effective in 1998 (Law Bridge, 1997-11-1) to safeguard occupants' interests such as contractor qualifications, construction project procedures, and building quality, among other criteria. However, enforcing the law seems to be difficult in the Beijing cases. Thus, improving construction quality remains a major challenge facing China's building industry.

Harmony with Humans: Matters of Social Cohesion

According to traditional Chinese philosophy, a residential environment should help regulate human behavior to create social harmony (Chapter 3), in modern words, social cohesion. The indicators here include education, occupation, and house-purchasing power, social relations among neighbors, relations with foreign neighbors, and indoor-outdoor visual interactions.

Education, Occupation, and House-Purchasing Power

Traditionally in China, girls would marry out to live with their in-laws and the joint family structure would require all family members to pool their income, a situation conducive to establishing family enterprise (Chapter 5).

This study shows that education *directly and indirectly* increases house-purchasing power. The interview results show that 63 percent of 40 residents in Beijing and Suzhou felt that education helps them to purchase a house and nurture good social relations important for employment or a rise in position. Economic reforms in China have made formal education more important for getting a better job and for earning more income than in the past. The pooled income from a couple and children also helps with buying a house, suggesting the financial strength afforded by family unity as in the past (Chapter 9).

However, four of 13 interviewed residents in Beijing's renewed traditional courtyard houses revealed that they remain where they are because they cannot afford to buy a new apartment elsewhere in the city. This finding confirms that of Zhen Li (2007).

Social Relations among Neighbors

Classical Chinese courtyard houses hosted single-extended-family members and these private homes became multi-household nuclear family compounds due to historic context (Chapter 1). Nevertheless, the social relations among neighbors were generally harmonious from the 1950s to the 1980s (Chapters 1 and 5).

This study shows that China's changing and polarizing society has a negative impact on social relations in Beijing and Suzhou. When redevelopments and reallocation replace much of the original neighborhood structure, gentrification leads to significantly weakened neighborly communications. This finding confirms those of Jin *et al.* (2004), Li (1998), Li and Wu (2006), Lu and He (2004), Tan (1994), and Wu (1999, pp. 172, 190). Nevertheless, from a planning perspective, dispersing a concentrated population from the city center is inevitable to alleviate such issues as congestion from overcrowding workers and traffic.

Activities between buildings are thought to be important for forming social relations, and seen as the necessary foundation for building a community. It is the arrangement of spatial elements, paths, and green areas that provides the opportunities for social interaction (Hargreaves and Webster, 2000, p. 6). The results show that communal courtyards can play a vital role in facilitating social relations in Beijing and Suzhou. A majority (64%; n=290) of the survey respondents considered their *courtyard* as the most important space for social relations. Suzhou residents considered a public corridor (33%; n=123) as their second most important place that fosters social relations, whereas for Beijing residents that venue is the *hutong* (25%; n=167), a finding consistent with that of Zhao (2000). Unlike English people who use pubs and markets as popular social spaces (Dines *et al.*, 2006; Watson and Studdert, 2006), only 4 percent of surveyed Beijing residents and 3 percent Suzhou residents regarded their local pub/store as a place for enhancing social relations.

Reduced courtyard size deteriorates social relations. The renewal of Beijing's traditional courtyard houses has not solved the problem of small living spaces, and communal courtyards in many compounds have become even smaller, regularly sparking disputes among neighbors.

Communal courtyards may increase the likelihood of neighborly encounters, and aid social relations more than other housing forms. The Nanchizi new courtyard housing and the larger new courtyards at Juer Hutong generally facilitate better neighborly communications and social relations than the three Suzhou cases, possibly due to the form and space design of the courtyards, among other factors.

Many residents observed that traditional multi-household courtyard compounds fostered better social relations than the new courtyard housing, and they still held fond, nostalgic memories of kind neighborly relations from the 1950s to the 1980s.

Fast-paced, modern lifestyles reduce the chances of neighborly communications because many people finish work late, and when they come home, their neighbors have gone to sleep. Modern facilities such as private bathrooms and kitchens also decrease the likelihood of neighborly communication.

There is a concern about using the communal courtyard for social events because of the noise and air pollution generated by them. As social activities in communal courtyards are not shared by all residents, especially where socio-cultural differences are perceived, these perceptions may cause resentment or tension among neighbors. Consequently, street gardens in Beijing are now important places for social and cultural activities because they are more public and accessible (Chapter 9).

Alternatively, planned communal activities help strengthen social relations. In the cases of Beijing's renewed traditional courtyard houses, Nanchizi new courtyard housing, and Suzhou Jiaanbieyuan, the local communities organize regular classes for senior citizens to learn about health, foreign languages, culture, science and technology, sports, and so on, which in turn contribute to enhanced social contacts after class.

Relations with Foreign Neighbors

Beijing Nanchizi and Juer Hutong new courtyard housing have attracted an influx of foreign residents (mostly from Europe and the United States) due to these projects' fame, convenient locations, and their 'Chinese' styles. The social relations between Chinese and non-Chinese neighbors are generally friendly, but remain at a superficial level due to language barriers, and differences in cultural backgrounds and lifestyles. In the eyes of some foreign residents, their Chinese neighbors are not very sociable. There have also been issues with foreign residents using the communal courtyard for noisy social events that disrupt and irritate Chinese residents.

Indoor-Outdoor Visual Interactions

Indoor-outdoor visual interaction through windows is a unique feature of courtyard housing where adults can supervise their children's activities in the courtyard while attending to their own chores. The survey shows that 54 percent of 290 respondents whose window faces a courtyard watch and take care of children's play through a window daily, and 37 percent do so weekly. In general, they spend an average of 30 minutes each time watching the children in the courtyard. However, the interview results reveal that although some residents enjoy these visual interactions, others have a less rewarding experience due to noise and occasional damage caused by them.

Discussion

The communal courtyards generally foster better social interactions and neighborly relations than a modern apartment building, but not as good as multifamily traditional courtyard compounds of the past. However, the decreased courtyard sizes in Beijing's renewed traditional courtyard houses have caused conflicts among neighbors, which contradict the doctrine of 'harmony with humans' in Chinese philosophy. The study suggests that neighborly relations are partly influenced by the form and space of courtyard housing, and partly by a changing and polarizing society shaped by residents' socio-economic levels, housing tenure, modern lifestyles, community involvement, common language, cultural awareness, and cultural background.

Harmony with Self: Time and Cultural Activities

Traditional Chinese philosophy indicates that a well-designed courtyard should facilitate self-cultivation (Chapter 3), and allow oneself to fully engage in nature's cycle by observing the fluctuations of time – the birth, growth, decay, death, and rebirth of natural elements (Al-Masri, 2010, p. 208; Mitchell, 2010, p. 234).

Ralph L. Knowles (1998, 1999) has noted that time is a result of daily and seasonal rhythms of sunlight, and sunlight adds a dimension of time to our perceptions of space. His research using a heliodon, a computer, or the outdoors under an open sky all point to the same findings, that any architectural space oriented north-south intensifies our experience of a day, and any space oriented east-west reinforces our experience of the seasons. In other words, if a wall faces east-west, it will highlight a daily rhythm and shadows will appear first on the west and then on the east, regardless of season. If a wall faces north-south, it will heighten a seasonal rhythm and shadows will expand much farther north in the winter than in the summer, regardless of the time of day. The crossing of spaces emphasizes multifaceted and contrapuntal rhythms of sunlight. Hence, Knowles (1998, 1999, p. 1) concludes:

> The result of the sun's daily passage is a shift in the patterns of dark and light in the great courtyards: west to east by day; north-south by year. In the shifting spot-lighted areas, it is easy to imagine ceremonial dances taking place at different locations in the courts at different times of the day and year.

Thus, a courtyard may offer people a unique experience of space and time which may uplift one to a spiritual realm.

Philosophy and Religion

There were generally three schools of thought in traditional China: Confucianism, Daoism, and Buddhism. The first two faiths originated in China, and the last one, India (Ching, 1993; Kohn, 2008; Qiao, 2002). The amalgamation of the three philosophical ideas is Neo-Confucianism (Kuwako, 1998; Weller and Bol, 1998). However, Communism has made a major shift in contemporary Chinese people's belief systems.

This study shows that 77 percent of 82 interviewed residents in Beijing and Suzhou have *no* philosophy or religion; although a small number of them (23%) have shown interest in learning traditional Chinese philosophy. Six percent (6%) are fond of Confucianism, Daoism, or Neo-Confucianism, and 17 percent prefer to follow Buddhist doctrines either by placing a Buddha statue at home, or paying respect to a Buddhist temple regularly. Only one resident referred to *Yi Jing* (The Book of Changes), but *Feng Shui* is often perceived as common sense and practical knowledge by residents who seem to express an embedded awareness of it in their survey answers, even though they seldom mentioned the word.

Cultural Activities at Home

For women, traditional cultural activities in Chinese courtyard houses would have included weaving, embroidering, sewing, educating children, and looking after senior family members, whereas for men, it might involve reading and reciting classics, writing essays or poems, practising calligraphy, painting, and so on (Chapter 5).

However, it is evident in these case studies that modern lifestyles have largely changed people's behavioral patterns. The survey and interview results show that besides sleeping, two major activities that Beijing and Suzhou residents do at home are watching TV/movies and working/playing on computer/browsing internet. This finding confirms that of Lu and He (2004) that watching TV has become a major daily activity.

The study also reveals that the living room is used most often for cultural activities, though bedrooms are used longer for sleeping, resting, or even working (the home as one's office in the case of Nanchizi).

Cultural Activities in Courtyards

Traditional Chinese courtyard houses fostered a variety of cultural activities such as playing with children, raising pets, playing chess, drinking tea, gardening, cooking in the summer, and so on (Chapter 5). What about the renewed/new courtyard housing? The survey of 290 residents shows that they spend most of their time indoors. Nevertheless, the top five cultural activities in the courtyards are (1) maintaining health/natural healing, (2) drinking tea, (3) gardening, (4) playing games, and (5) holding birthday parties. However, some cultural activities such as composing poetry or essays, practicing calligraphy, dancing, playing musical instruments, painting, holding a wedding ceremony, or singing traditional opera are much less frequent or no longer performed in the courtyards, possibly because modern lifestyles no longer support such activities.

Time is a constraint to conducting cultural activities in courtyards. The survey shows that residents normally hold their *favorite* cultural activities in courtyards after regular work hours. Moreover, cultural activities in courtyards are mostly performed during weekends. Despite a time constraint, most residents (53%;

n=201) use their courtyard at least 1-3 times weekly, and some use it at least 1-3 times monthly (31%).

Climate seems to be another compelling factor that affects residents' cultural activities in courtyards. Suzhou residents (n=123) were generally noted to have a significantly higher degree of courtyard uses for cultural activities than their Beijing counterparts (n=167), when comparing their mean average values and percentages of the time of day, length of time, day of the week, frequency, and seasons they spend in courtyards. One reason seems to be that Suzhou has a warmer climate (average annual temperature 15°-17°c) than that of Beijing (average annual temperature 10°-12°c) (China.org.cn, n.d.).

For Beijing residents, summer is their most active season, followed by spring and autumn because they are serene, and winter is their least active season due to severe, cold weather. For Suzhou residents, autumn and spring are their most active seasons whereas summer is not as it is too hot. Unlike Beijing residents, there is no dramatic decline in Suzhou residents' winter cultural activities since the weather is relatively mild.

Courtyard ownership may also affect residents' cultural activities. The findings indicate that Suzhou residents conduct more cultural activities in courtyards than their Beijing counterparts, possibly because Suzhou residents (38%; n=123) have significantly more private courtyards than Beijing ones (0.6%; n=167). Moreover, south-facing balconies in Beijing Juer Hutong new courtyard housing have supported cultural activities, much like dew platforms in classical Chinese courtyard houses in the past (Chapter 5).

Courtyard size is important to cultural activities as across all the cases, residents complained about their small courtyard sizes that discourage them from lingering there. Moreover, a lack of courtyard recreational facilities also prevents their use. For example, if stone tables and stools were placed in the courtyard, elderly residents could sit, drink tea, or play games, although these activities may also generate unwanted noise.

Cultural Activities in Community/City Parks/Gardens

Traditionally, Chinese women would rarely step outside their courtyard walls to conduct cultural activities except for special occasions (Chapter 5). This study shows, however, community/city parks/gardens have become important alternatives for daily cultural activities because they are more public, although they are not as convenient as a courtyard at home. Significantly more Beijing residents (75%; n=167) spend their time in public outdoor spaces than Suzhou ones (47%; n=123). Thus, regional cultural differences seem to have an effect on the use of public outdoor spaces.

Cultural Festivities Indoors and Outdoors

Classical Chinese courtyard houses could accommodate *all* the Han Chinese cultural festivals celebrated in their full scale and detail (Chapter 5). This study shows, nevertheless, only major cultural festivals related to seasonal change are currently celebrated: Spring Festival/Lunar New Year, Dragon Boat Festival (originally related to Summer Solstice), Mid-Autumn Festival, and Winter Solstice Festival, along with festivals related to spiritual/ancestral worship such as Lantern Festival (or 'Spring Spirit Festival') and Qing Ming Festival (or 'Tomb-Sweeping Day'). Certain festivals such as Blue Dragon Festival and Mid-Summer Spirit Festival are not celebrated as much anymore, possibly due to a lack of private courtyard space to support these festivities.

The survey of 290 respondents shows only three *favorite* cultural festivities conducted in the courtyards: enjoying the full moon on the night of Mid-Autumn Festival (1.4%), letting off firecrackers/fireworks on the night of Lunar New Year's Eve/Spring Festival (0.7%), and watching lantern displays on the night of Lantern Festival (0.3%). These findings imply that most residents celebrate cultural festivities indoors, perhaps due to cold weather at Spring Festival and insufficient private courtyard for Mid-Autumn Festival. Nevertheless, in both Beijing and Suzhou, some residents go to public places in the city to enjoy the view of the full moon on Mid-Autumn Festival. Many festivities remain almost traditional (Chapter 5), although they may be blended with modern rituals such as a restaurant meal, karaoke at a night club, or simply watching festival galas on TV at home.

Four of 82 interviewed residents revealed that decorations for Spring Festival are for their children/grandchildren. If their children/grandchildren are away from home, they would feel unmotivated to decorate. The research also indicates that stem family tradition is preserved in a few households where three generations live together. At the same time, the smallness of most housing units makes stem family living very difficult.

Birthday Celebrations Indoors and Outdoors

Traditional birthday celebrations in China were mainly for children and senior family members, mostly conducted in the Central Hall, with celebrations occasionally spread out to the courtyard (Chapter 5). This survey (n=290) shows that holding a birthday party is the 5th major cultural activity in the courtyards, and 47 percent of 70 interviewed residents celebrate their birthdays. These celebrations remain somewhat traditional with modern touches such as a restaurant meal or karaoke at a night club. In one courtyard compound at Nanchizi new courtyard housing, residents sang 'Happy Birthday to You' to a neighbor, revealing a degree of social cohesion created by the communal courtyard for this occasion.

Wedding Ceremony/Anniversary Celebrations

A traditional Chinese wedding ceremony would be conducted in the courtyard (or Central Hall) at home because the courtyard created a direct link between Heaven and Earth to which the bride and groom would pay respect in their wedding vows (Chapter 5). The survey of 290 residents in Beijing and Suzhou shows a low mean average value of holding a wedding ceremony in the renewed/new courtyards, and none of the residents considered wedding as their *favorite* cultural activity in the courtyards. The fact that weddings are rare activities may well explain why people do not rank it as a favorite ritual. There may also be few residents who actually got married during their residencies in these housing estates.

Traditionally in China, people would not celebrate wedding anniversaries. Nonetheless, 27 percent of 70 interviewed residents celebrate wedding anniversaries nowadays to various degrees, showing Western cultural influence. The celebrations often include a meal in a romantic place, the exchange of gifts at home, or a trip to a scenic spot. Residents in Beijing and Suzhou have adopted the Western notions of 'silver (25th),' 'gold (50th),' and 'diamond (60th)' for their wedding anniversaries celebrations.

Discussion

These case projects suggest that the renewed/new communal courtyards only facilitate harmony with self or self-cultivation to various extents, but not as much as the single-family private courtyards of the past. The factors that affect their cultural activities in courtyards include modern lifestyles, time, climate, courtyard ownership, yard size, facilities, and so on.

Summary and Conclusion

This chapter synthesized the findings and attempted to answer the main research questions. The results show that the renewed/new courtyard housing is only culturally sustainable to various degrees and in different contexts; they have not achieved this harmonious state of being due to the multitude of issues mentioned in the chapter.

The chapter proposed four cornerstones of culturally sustainable architecture in China: (1) harmony with heaven: form and environmental quality; (2) harmony with earth: space and construction quality; (3) harmony with humans: matters of social cohesion; and (4) harmony with self: time and cultural activities. This framework of measurement (table 11.1) may have implications for other cultures where courtyard housing was, or still is, their traditional living patterns.

This study reveals that Beijing's renewed traditional courtyard houses are largely unsuccessful because the renewal has not entirely complied with the principles prescribed by the municipal government, and that the contractors did not possess essential knowledge or skills for undertaking serious construction projects. However, it does not necessarily suggest that historic conservation is an inappropriate approach in other social contexts. Similarly, although numerous issues mentioned by residents about new courtyard housing experiments in Beijing and Suzhou signify that the designs and/or the constructions need improvement; they do not indicate that the courtyard concept is flawed. Also, some difficulties are obviously related to political, economic, and social factors rather than architectural designs.

In Chapter 12 that follows, the conclusions, contributions to the tree of knowledge, and suggestions for creating new courtyard garden houses, will be presented to highlight the implications of the study.

Table 11.1   Four cornerstones of culturally sustainable architecture in China

Material/Dwelling Culture (archi-culture, tangible)		Immaterial/Spiritual Culture (socio-culture, intangible)	
Harmony with heaven: form and environmental quality	Harmony with earth: space and construction quality	Harmony with humans: matters of social cohesion	Harmony with self: time and cultural activities
• Exterior form • Exterior walls • Gate and access • Windows • Courtyards and gardens • Roofs	• Interior space • Floor levels • Furniture styles and materials • Facility provision • Building materials and construction quality • Maintenance and management • Car park spaces	• Education, occupation, and house-purchasing power • Social relations among neighbors • Relations with foreign neighbors • Indoor-outdoor visual interactions	• Philosophy and religion • Cultural activities at home • Cultural activities in courtyards • Cultural activities in community/city parks/gardens • Cultural festivities indoors and outdoors • Birthday celebrations • Wedding ceremony or anniversary celebrations

Source: My summary

PART V
Future New Courtyard Garden Houses

Chapter 12
Conclusion: In Search of Paradise[1]

If Heaven intends culture to be destroyed,
those who came after me will not be able to have any part of it.
If Heaven does not intend this culture to be destroyed,
then what can the people of Kuang do to me?
— Confucius, 551-479 BCE, *Analects*, Book 9 Chapter 5

This chapter answers the research questions and addresses the research objectives. It consists of three sections: holistic view – health and courtyard; contribution to the tree of knowledge; and creation of new courtyard garden houses.

Holistic View: Health and Courtyard

This was a collective case study of six renewed/new courtyard housing estates built in Beijing and Suzhou since the 1990s. It adopted a social research methodology of qualitative and quantitative inquiries mainly using field surveys, interviews, time diaries, and historical texts to measure the cultural sustainability of these projects. This topic is important because the decline of traditional courtyard houses and the issues associated with new housing forms all point to the necessity to investigate the new courtyard housing for better housing design and development in the future.

The findings indicate that certain features in classical Chinese courtyard houses are still valued by residents in renewed/new courtyard housing in Beijing and Suzhou, such as the courtyard, low-rise, southeast/south gate orientation, south-facing window orientation, north-south cross ventilation orientation, and so on. The new communal courtyards still can facilitate neighborly communications and social relations, though in a different way and not as much as traditional courtyards of the past. The results suggest that neighborly relations are only partly influenced by the form and space of the courtyard housing and are perhaps influenced even more so by China's changing and polarizing society as manifested in these specific residents' socio-economic levels, housing tenure, modern lifestyles, community involvement, common language, cultural awareness, and demographic backgrounds.

---

1   The word 'paradise' here comes from a Chinese phrase: "In Heaven, there is Paradise; on Earth, there are Suzhou and Hangzhou" (上有天堂, 下有苏杭), suggesting that the courtyard gardens in these two cities represent the earthly paradise.

The research also shows that the new courtyard housing facilitates some traditional cultural activities, the primary one being maintaining health/natural healing, along with drinking tea, gardening, playing games for recreation, and holding birthday parties, all of which contribute to occupants' physical, mental, and emotional health and wellbeing. Nevertheless, many other traditional cultural activities are much less often or no longer performed in the communal courtyards, likely due to factors such as time constraints, climate, courtyard ownership, yard size, facilities, community involvement, and so on.

Is the courtyard really essential for sustaining Chinese culture? In principle, a courtyard helps residents to connect with nature, with other humans, and with themselves, which, in Chinese philosophy, is important for healthy living in the past and present. Although difficult to prove, it is widely accepted that there is a close relationship between human health and nature (Hartig and Marcus, 2006; Lyle, 1994, p. 100; Marcus, 2004; Ulrich, 1984). Courtyards offer light, air, and views of nature, where trees, plants, and flowers can grow. Plants in courtyards are natural processors of air that absorb carbon dioxide, emit oxygen and water vapor, and take in pollutants through their leaves. If properly placed, courtyard trees with a dense canopy of leaves have significant cooling effects that create pleasant microclimates (Kou, 2005, pp. 31, 82; Lyle, 1994, pp. 26, 102-104, 116). All these factors are conducive to the health and wellbeing of the occupants of courtyard housing. However, due to high population density and land constraints in China, to aim for the optimum courtyard size for each household may remain a classical ideal and a challenge for contemporary Chinese planners and architects to achieve.

The findings suggest that the renewed/new courtyard housing projects are only culturally sustainable to various degrees and in different contexts; they have not achieved this harmonious state of being partly due to the form and space design, and partly due to the poor quality of building materials and construction, particularly in the Beijing cases. These problems cause Beijing residents more concern than the enjoyment they gain from their courtyards.

Contribution to the Tree of Knowledge

The contributions of this study to new knowledge include most of the empirical findings, and the theorization of cultural sustainability and culturally sustainable architecture in China. Traditionally, cultural sustainability had been regarded as part of social sustainability and thus is less researched due to the difficulty of measuring it. This study has attempted to separate the two dimensions although with occasional overlaps.

Previous studies indicate that cultural sustainability encompasses concepts such as cultural vitality, cultural diversity, and cultural activities because cultural activities (including festivities) function as catalysts for cultural sustainability. The renewed/new courtyard housing in Beijing and Suzhou facilitates traditional Chinese cultural expressions to varied degrees. Thus, we may say that they help sustain traditional Chinese culture to some extent. Nevertheless, cultural activities and festivities in courtyards seem to be a higher pursuit once basic housing functions have been met, but this has not always been the case in Beijing. Hence, the sustainability of architectural culture or archi-culture is a prerequisite of the sustainability of socio-culture in China.

We now know that cultural sustainability encompasses the sustainability of both material and immaterial/spiritual cultures. The cultural sustainability indicators are the four key themes identified in Chinese philosophy: harmony with heaven, harmony with earth, harmony with humans, and harmony with self, along with form, space, matter, and time. These four aspects comprise the four cornerstones of culturally sustainable architecture in China for measuring new housing design projects. More indicators may be added in future studies.

Symbolically, courtyard housing is an ideal place to work out cultural sustainability because culture has been regarded as the fourth pillar of sustainable development alongside the other three: environmental responsibility, economic viability, and social equity. The concept of four-sided enclosure strongly present in courtyard housing corresponds to the four pillars of sustainable development. Moreover, the courtyard at the center signifies political support from the central government. This study indicates that Chinese planning policies and national building codes have both endorsed the continuity of courtyard form and meanwhile, have restricted the architectural designs, leaving only few options for the architects to achieve more ideal schemes. Hence, political support should be the additional *fifth* pillar of sustainable development, without which other sustainability will be difficult to accomplish, and which should be investigated further.

As courtyard living complies with Chinese philosophy, which emphasizes harmony, it also requires that all things in the world reciprocally support and sustain one another. Harmony is conducive to nurturing renewed creativity and the future development of humankind because the continuous existence and evolution of the world's civilizations depend on the harmonious relationship of all things. Hence, the ultimate purpose, in traditional Chinese philosophy, of living in harmony with the natural and cultural environments, is to be healthy, to obtain the unity of heaven and humans, and to reach a sense of cosmic oneness (Berthrong, 1998, p. 259; Capra, 1975/1999, p. 326; Cheng, 1998, p. 219; Kohn, 2001, pp. 389-390; Ro, 1998; Tucker, 1998; Zhang, 2001, p. 366).

Creation of New Courtyard Garden Houses

A square lǐ covers nine squares of land,
which nine squares contain nine hundred mu.
The central square is the public field,
and eight families, each having its private hundred mu,
cultivate in common the public field.
And not till the public work is finished,

may they presume to attend to their private affairs...
Those are the great outlines of the system.
In the fields of a district,
those who belong to the same nine squares
render all friendly offices to one another
in their going out and coming in,
aid one another in keeping watch and ward,
and sustain one another in sickness.
Thus the people are brought to live in affection and harmony.
– Mengzi, 372-289 BCE, *Works*, Book 3 Part 1 Chapter 3

Imperial Chinese capital cities and classical courtyard houses were walled enclosures, whose planning and design complied with Chinese philosophy. An outstanding example is Tang (618-907) Chang'an (now Xi'an) with 108 urban blocks (110 including the markets), each was a ward (坊 *fang* or 里坊 *lifang*) enclosed by earthen walls of various sizes, most of which were residential (Fu, 2002b; Liang, 1998, p. 292). A typical residential ward had 4-16 internal divisions, though officials and aristocrats had relatively larger plots. The walled compounds acted as small fortresses and indicators of a clan's strength and power (Fu, 2002b; Knapp, 2005a).

The world is forever in a cyclic process of 'creation-destruction-recreation' (Eliade, 1959, p. 108), and the massive dismantling of China's traditional courtyard houses in fact offers opportunities to create new ones in harmony with the old. However, this is not arguing for a simple return to old lifestyles or copying classical courtyard house designs in contemporary China, but suggesting ways to create new courtyard garden houses in the light of current material, technological, economic, social, and cultural development.

Exterior Form

To be culturally sustainable, the form of new courtyard garden houses should follow the 'Nine Squares' or 'Four Squares' traditional Chinese planning principle, with a clear demarcation of semi-public and private outdoor space for both social and solitary activities because this research shows that communal courtyards/ gardens are conducive to social interaction and private courtyards/gardens to self-cultivation, and that the courtyard size should be proportional to the height of its surrounding buildings (minimum 1:3 for Beijing and 1:1.3 for Suzhou) to allow sufficient sunlight. This research also suggests that small, enclosed compounds without automobiles promote social cohesion and cultural activities more than large housing estates.

In the Beijing scheme, each compound measures 60 m × 60 m from the enclosing walls, which can be used as back walls for a 'commercial strip' or 'shopping strip' (商业街), on the basis that the residents' privacy is protected, and that the air quality and noise level are well controlled. In Beijing, a typical *hutong*

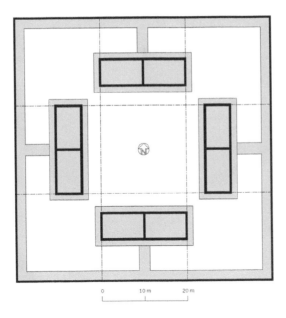

Figure 12.1 Proposed Beijing new courtyard garden house (北京新四合园) system based on the 'Nine Squares' system. Design and drawing by Donia Zhang 2008-2012

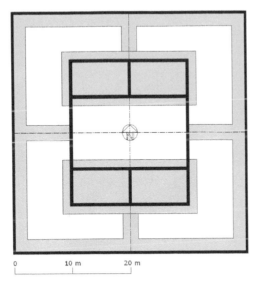

Figure 12.2 Proposed Suzhou new courtyard garden house (苏州新回田园) system based on the 'Four Squares' system. Design and drawing by Donia Zhang 2008-2012

Figure 12.3   Model of the proposed Beijing new courtyard garden house compound housing 8 nuclear families. Design and model by Donia Zhang 2008-2012

Figure 12.4   Model of the proposed Suzhou new courtyard garden house compound housing 4 nuclear families. Design and model by Donia Zhang 2008-2012

('lane') block distance is 70-80 m, and a typical plot 73 m × 60 m or 77 m × 63 m (Wang, 1999, p. 23; Wu, 1999, p. 77). Hence, the 60 m × 60 m standardized compound can comfortably fit into the existing *hutong* structures and be modified according to each site conditions.

A typical *hutong* length of 440 m may accommodate seven such compounds, accounting for 56 households (each compound for eight households), as opposed to only seven households in classical courtyard houses in a *hutong*, albeit the single-extended-family size was considerably larger then. Like living in traditional courtyard houses, residents in these small compounds can easily walk to streets, shops, bus stops, and other public amenities, meanwhile, they will have a strong sense of territoriality because they will know their neighbors.

In the Suzhou scheme, each compound measures 40 m × 40 m from the enclosing walls, which can also be used as back walls for commercial buildings. A typical block distance between the *nong* ('lanes') is 60-80 m (Chen and Romice, 2009; Yu, 2007, p. 13) that can easily fit for such a compound, and a typical *nong* length of 200-400 m may accommodate 5-10 compounds or 20-40 households.

However, it is important to avoid creating concentrations of housing of the same type in one particular area, and a mixed-use neighborhood is more robust and sustainable than a mono-residential development (Carmona *et al.*, 2003; Hester, 2006).

Exterior Walls

The enclosing walls of traditional courtyard houses are still a valued feature in contemporary Beijing, especially for children (Zhao, 2000). Professor Zhu Wenyi at Tsinghua University has analyzed the historical and cultural connotations of enclosing walls in ancient China, and argues that the walls were to create spatial boundary of three-dimensional spaces, since without them humans can only experience two dimensions on the earth plane. The walls of humble, one-family courtyard houses and the entire *hutong*, and those of palaces, cities, and even the Great Wall of China, all reflected people's desire to live in protective, three-dimensional spaces. Thus, the ideal living environment for the ancient Chinese was spaces enclosed entirely by walls, with gates only for movement between the spaces. Nevertheless, the people wanted to make up the 'defect' (the Chinese considered an opening or hole as incomplete) by patching it with screens, secondary walls, and so on (Beijing Culture Net, 2002).

The proposed 3-meter-high enclosing walls for Beijing and Suzhou new courtyard garden houses have lattice/assorted windows facing the lanes/streets for pedestrians to have interesting views of the gardens inside.

Gates and Access

Since residents still prefer traditional gate orientations, the Beijing scheme has two main gates at the southeast and northwest corners, each installed with mailboxes

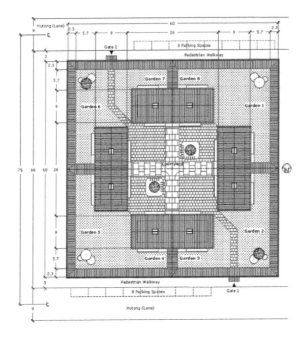

Figure 12.5   Beijing new courtyard garden house compound site plan. Design and drawing by Donia Zhang 2008-2012

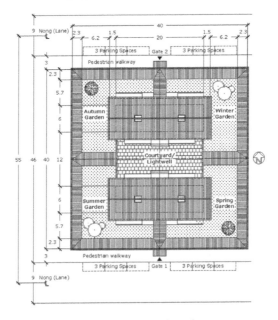

Figure 12.6   Suzhou new courtyard garden house compound site plan. Design and drawing by Donia Zhang 2008-2012

and an intercom system, and two side gates as emergency exits or garden connections with the adjacent compounds. The Suzhou scheme has two main gates directly facing the south and north directions, each mounted with mailboxes and an intercom system, and two side gates as emergency exits or garden connections with the adjacent compounds. Each housing unit has verandas, balconies, and barrier-free accesses along the edges of the building. The verandas encourage walks, exercise, art display, and social interaction, especially on rainy and snowy days, and protect residents from summer sunburn. Urban design research findings show that popular zones for staying are along the facades in a space or in the transitional areas between one space and another where one is able to view both spaces at the same time but without being seen too much, because when one's back is protected, it feels safer. The most natural place to linger is the doorstep where one can go further out into the space or remain standing (Alexander *et al.*, 1977; Gehl, 1971/2001, pp. 151-152; Hall, 1966).

Windows

Because residents prefer traditional window orientations for better sunlight, natural ventilation, and views of nature, each housing unit has been designed with a skylight and windows facing three cardinal directions.

Figure 12.7 Beijing new courtyard garden house elevations facing the gardens. Design and drawing by Donia Zhang 2008-2012

Figure 12.8 Beijing new courtyard garden house elevations facing the courtyard. Design and drawing by Donia Zhang 2008-2012

Figure 12.9   Suzhou new courtyard garden house elevations facing the courtyard. Design and drawing by Donia Zhang 2008-2012

Figure 12.10 Suzhou new courtyard garden house elevations facing the gardens. Design and drawing by Donia Zhang 2008-2012

Courtyards and Gardens

Since communal courtyards/gardens facilitate social interaction and private courtyards/gardens foster self-cultivation, providing both of them in the immediate reach of each household is best. With the economic boom in recent China, gardens have once again become fashionable (Meyer, 2001, p. 233). The garden designs should follow the principles exhibited in classical Suzhou gardens because Chinese people traditionally believed that the composition of rocks and water, trees and grass has symbolic meanings and significantly affects the development of human body, mind, and spirit (Wang, 2005, p. 76).

In his memoir, *Six Records of a Floating Life* (浮生六记), Shen Fu (1763-1810?) outlines the philosophy of Chinese garden design:

1. Creating the large (*yang*) in the small (*yin*) and the small in the large;
2. Providing the real (*shi*) in the unreal (*xu*) and the unreal in the real;
3. Using contrastive views such as hidden and obvious, shallow and deep, and near and far;

4. Making plants the main views, less is more (Wang, 2005, p. 83; Yu, 2007, pp. 245-248).

Yu Shengfang (2007) offers an additional set of Chinese garden design guidelines:

1. Building pavilions and corridors along the perimeter of the garden to maximize the effective use of space;
2. Planting trees near the sides or corners of the garden to not block the views;
3. Placing rocks and flower beds against the enclosing walls to save space and having the walls as the backdrop for reflections;
4. Adding a water feature and making the water safe for children and the elderly by changing its level from shallow to deep;
5. Creating a focal point for all the scenic elements viewed from the Central Hall to form a complete picture in front of the room; and
6. Designing the elements so that they can be viewed closely because the garden is small and the viewing distance is short, covering 180° angle for seeing from left, middle, and right (pp. 244, 337).

In the Beijing scheme, the four buildings are arranged in direct east, west, south, and north orientations forming a communal courtyard of 26 m × 26 m shared by eight nuclear families, with the private gardens at the back for each household's cultural activities. The courtyard size complies with my previous research finding that if it is to achieve the same amount of sunlight as in a traditional Beijing *siheyuan*, the ratio of building height to distance should be minimum 1:3 (Zhang, 1994, 2006, 2009/2010/2011).

In the Suzhou scheme, the two buildings face the south and north orientations with a 3-meter-high garden wall pierced by a moon gate in the east and west directions, forming a communal courtyard of 12 m × 20 m shared by four nuclear families (because east- or west-facing buildings are not Suzhou's tradition). The Suzhou communal courtyard is smaller than that of Beijing to reduce the admission of hot summer sunlight, prevent heatstroke and lower the temperature. It is meant to be reminiscent of the traditional Suzhou lightwell. The 12 m dimension has met Suzhou's planning regulation of the ratio of building height to distance of 1:1.3. The themed gardens (spring, summer, autumn, and winter) reflecting the four seasons are designed around each housing unit for cultural activities.

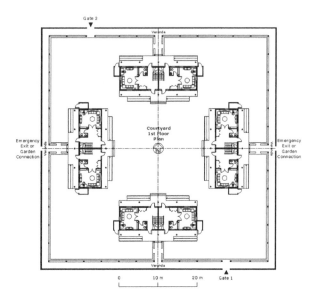

**Figure 12.11 Beijing new courtyard garden house compound 1st floor plan.
Design and drawing by Donia Zhang 2008-2012**

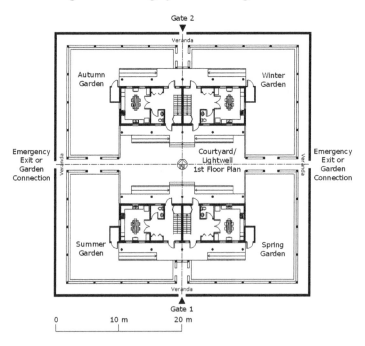

**Figure 12.12 Suzhou new courtyard garden house compound 1st floor plan.
Design and drawing by Donia Zhang 2008-2012**

Roofs

As residents prefer pitched roofs than flat ones for their better thermal performances, pitched roofs at a slope of 6:12 are designed for both the Beijing and Suzhou schemes. The grey tiles for Beijing and black tiles for Suzhou respect their local traditions. A skylight is designed for each housing unit for admitting an even amount of daylight. Eaves are designed 1.8 m deep to shelter the balconies (1.5 m deep) and prevent rainwater from slanting in.

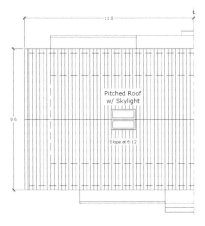

Figure 12.13 Beijing new courtyard garden house roof plan. Design and drawing by Donia Zhang 2008-2012

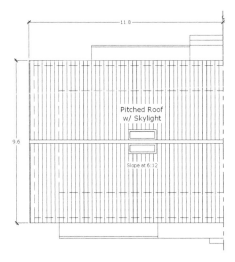

Figure 12.14 Suzhou new courtyard garden house roof plan. Design and drawing by Donia Zhang 2008-2012

Interior Space

To be culturally sustainable, the interior space of new courtyard garden houses should be proportional to each household size for various cultural activities and festivities.

In both the Beijing and Suzhou schemes, each housing unit is rectangular in plan, with a width of 10 m, depth of 6 m, and a total internal floor area of 180 sqm. This wide frontage and shallow depth plan is advantageous for admitting sunlight into the rooms without the need for artificial lighting which allows for energy efficiency (Amin, King, and Zhang, 1994, p. 69; Hayward, 1994). It has also eliminated the disadvantage of having to travel outside from room to room even in severe weathers as in a traditional courtyard house. The 180 sqm unit size is popular according to Wu Fulong (2004) who observes that the best-selling housing type in Beijing has three bedrooms and two receptions at a size of 150-170 sqm (p. 231). Likewise, architect Wu Chen indicated that current housing unit designs in Beijing are 180 sqm (interview, 2008).

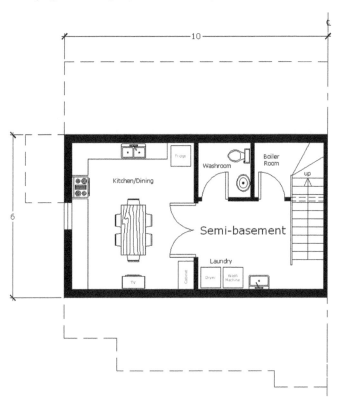

Figure 12.15 Beijing new courtyard garden house semi-basement plan. Design and drawing by Donia Zhang 2008-2012

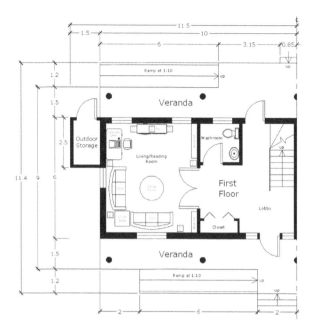

Figure 12.16 Beijing new courtyard garden house 1st floor plan. Design and drawing by Donia Zhang 2008-2012

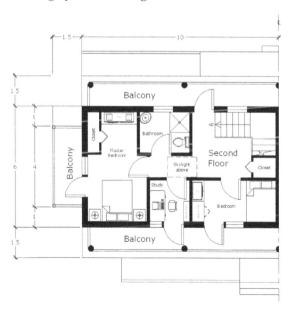

Figure 12.17 Beijing new courtyard garden house 2nd floor plan. Design and drawing by Donia Zhang 2008-2012

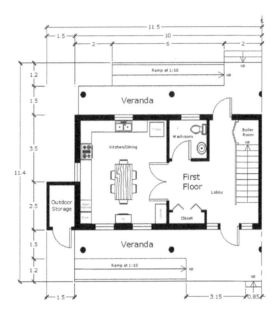

Figure 12.18 Suzhou new courtyard garden house 1st floor plan. Design and drawing by Donia Zhang 2008-2012

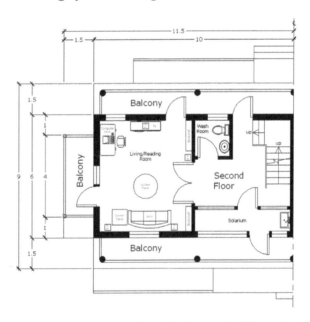

Figure 12.19 Suzhou new courtyard garden house 2nd floor plan. Design and drawing by Donia Zhang 2008-2012

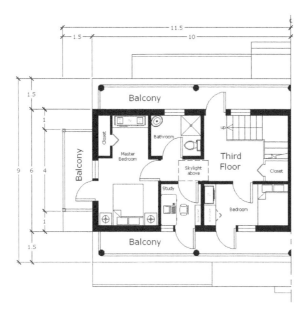

Figure 12.20 Suzhou new courtyard garden house 3rd floor plan. Design and drawing by Donia Zhang 2008-2012

Floor Levels

This research has revealed that residents prefer 1-3 storeys. Hence, Beijing new courtyard garden houses are designed with 2 ½ storeys (with a semi-basement), and Suzhou ones with 3-storeys because traditional Suzhou houses did not have basements due to its damp climate.

Facility Provision

Modern facilities such as a kitchen, bathroom(s), running water, electricity, gas, heating system, air-conditioning, TV and Internet cables, telephone lines, and so on, should be provided in each housing unit, and an outdoor storage of 2.5 m × 1.5 m to be attached to the side of each house for storing gardening tools, and recycling and garbage bins.

Bicycle racks, a lamp post, and a drainage hole should be installed in the communal courtyard.

Building Materials and Construction Quality

Since stone, brick, or concrete are more enduring and sustainable than wood, and they do not require cutting down trees, they should be applied as basic building materials. The party walls between two households should be constructed with a high level of acoustic insulation to reduce noise invasion. Good-quality construction is of paramount importance, against which the final housing product is measured.

Car Park Spaces

Determined by the compound dimensions, 16 car park spaces (200% provision), eight on each side of the lane are provided in the Beijing scheme, and 12 car park spaces (300% provision), six on each side of the lane in the Suzhou plan. These arrangements ensure the courtyards and gardens are safe places for children to play, and the elderly to conduct social and cultural activities as China has a large and growing aging population who have a considerable amount of daytime to spend with their neighbors.

Gehl (1971/2001, pp. 79, 113, 129) notes that leaving cars at the edge of residential areas and walking between 50-200 m home through the neighborhood has become increasingly common in new European residential areas. The farther away from the doors the cars are parked, the more activities will take place in the outdoor space.

The proposed pedestrian sidewalk width is 3 m, a good dimension for pedestrian traffic flow. In Venice, where no automobiles are permitted, the average street width is 3 m (Gehl, 1971/2001, p. 93).

Density and Plot Ratio

Assuming 3-5 persons in each household, the Beijing scheme has a density of 67-111 persons per hectare or 27-45 persons per acre and plot ratio of 1:1.5 (table 12.1), whereas the Suzhou scheme has a density of 75-125 persons per hectare or 30-50 persons per acre and plot ratio of 1:2.5 (table 12.2). However, the less quantifiable, social and cultural implications of the designs cannot be estimated. Regarding density, Amos Rapoport (1983) rigorously argues that density needs to be considered not in terms of persons per unit area, but in terms of perceived density, that is, the relation of buildings to spaces, and also in terms of crowding, desired level of involvement, unwanted interaction in all senses, and so on.

Table 12.1   Beijing new courtyard garden house density and plot ratio

Beijing New Courtyard Garden House		
Built floor areas "A":		
A_0 = 1396 sqm (verandas, balconies, storages, stairs)		
A_1 = 6×10×8 = 480 sqm (1st floor)		
A_2 = 6×10×8 = 480 sqm (2nd floor)		
$A = A_0 + A_1 + A_2$ = 2356 sqm (total)		
Number of households: 8		
Number of persons: 24-40 (3-5 persons per household)		
Site area = X × Y = 46.4×46.4-11.7×11.7×4 = 1605 sqm		
Plot area = 60×60 = 3600 sqm = 0.36 hectare = 0.9 acre		
Density =	Number of persons in block	
	Plot area	
Density = 67-111 persons/hectare = 27-45 persons/acre		
Plot ratio =	Site area	
	Built floor area	
Plot ratio = 1:1.5		

Note: 1 hectare = 2.47105 acre = 10,000 sqm; or 1 acre = 0.404685 hectare = 4046.8 sqm; basement does not count

Table 12.2   Suzhou new courtyard garden house density and plot ratio

Suzhou New Courtyard Garden House
Built floor areas "A":
$A_0 = 971$ sqm (verandas, balconies, storages, stairs)
$A_1 = 6\times10\times4 = 240$ sqm (1st floor)
$A_2 = 6\times10\times4 = 240$ sqm (2nd floor)
$A_3 = 6\times10\times4 = 240$ sqm (3nd floor)
$A = A_0 + A_1 + A_2 + A_3 = 1691$ sqm (total)
Number of households: 4
Number of persons: 12-20 (3-5 persons per household)
Site area = X × Y = 23×29.4 = 676 sqm
Plot area = 40×40 = 1600 sqm = 0.16 hectare = 0.4 acre

Density =	Number of persons in block
	Plot area

Density = 75-125 persons/hectare = 30-50 persons/acre

Plot ratio =	Site area
	Built floor area

Plot ratio = 1:2.5

Note: 1 hectare = 2.47105 acre = 10,000 sqm; or 1 acre = 0.404685 hectare = 4046.8 sqm

Table 12.3   Summary of the proposed new courtyard garden houses

Design Element	Beijing Scheme	Suzhou Scheme	Reasons
1. Exterior form	Using the 'Nine Squares' system; 60 m × 60 m standardized compound for 8 households	Using the 'Four Squares' system; 40 m × 40 m standardized compound for 4 households	Cultural tradition and standardization
2. Exterior/ Enclosing walls	White enclosing walls surrounding the compound with lattice/assorted windows	White enclosing walls surrounding the compound with lattice/assorted windows	Cultural tradition and safety concerns
3. Gates and access	Gates at traditional southeast and northwest corners; verandas, balconies, and barrier-free access	Gates at traditional south and north orientations; verandas, balconies, and barrier-free access	Cultural tradition, nature connection, and care for the disadvantaged groups
4. Windows	A skylight with windows facing 3 cardinal directions	A skylight with windows facing 3 cardinal directions	Sunlight, natural ventilation, and views of nature
5. Courtyards and gardens	26 m × 26 m courtyard with 8 private gardens	12 m × 20 m courtyard with 4 private gardens	Social interaction and self-cultivation
6. Roofs	Grey-tiled pitched roofs at a slope of 6:12	Black-tiled pitched roofs at a slope of 6:12	Cultural tradition and thermal comfort
7. Interior space	180 sqm	180 sqm	Spacious for a 2-5 person family
8. Floor levels	2 ½ storeys (with a semi-basement)	3 storeys	Residents' preference
9. Facility provision	Basic modern facilities	Basic modern facilities	Conveniences
10. Building materials and construction quality	Stone, bricks, and concrete materials constructed in high quality	Stone, bricks, and concrete materials constructed in high quality	Durability and sustainability
11. Car park spaces	16 (200% provision)	12 (300% provision)	Abundance even for visitors
12. Density and plot ratio	67-111 persons/ha; 1:1.5	75-125 persons/ha; 1:2.5	Spatial comfort

Source: My summary

Discussion

Since a house is a fairly permanent structure, once built, it cannot be changed easily to accommodate newer demands or higher standards. Therefore, housing designs should not be compromised for less than stable requirements in density, plot ratio, or floor-area ratio because while a population may fluctuate with time, a housing form may be less flexible. It is in fact more economical and environmental to build for the long-term than to demolish and rebuild at a later time.

Multiple new courtyard garden house compounds can be built side by side in traditional *hutong/nong* structures. This kind of communal living is important, philosophically and practically. Humans need to live in harmony with other humans, and only by doing so can we expect to live harmoniously with the nonhuman world. This capacity is especially crucial for China that is implementing the 'One Family One Child' policy because the single child is often self-centered who tends to show little concern for others. This may become a social issue when they enter society.

A perceived problem of the schemes is that the neighbors may have conflicts due to different work-life schedules that may result in unwanted noise. This drawback can be controlled by applying good acoustic insulation between the party walls. For the communal courtyard, a rule can be negotiated and established among the neighbors as to the activities, and their timing, that are allowed or not in the courtyard.

These two schemes are merely a suggestion – different interpretations of the guiding principles established in this section may result in different designs which may bring variety and diversity to a city and which may have implications for other places in the world.

Epilogue

Be the first to worry about the woes of the people,
and the last to share the weal of the people.

– Fan Zhongyan (范仲淹, 989-1052)

The main motivation for me in conducting this research is to care for the common people living in these Chinese cities, to reflect their hopes, aspirations, anxieties, and even anger when some of them talked about the poor-quality homes in which they live.

The Chinese government has been enthusiastically promoting the 'people-centered' (以人为本) principle in many of its sectors, and the words often appear in Chinese media today. However, putting this motto into action is often a more challenging task. In future, not only architectural designs should take the people-centered approach, our research goals and methods should also follow the people-centered path.

This piece of work feels like a major assignment in my life. As Zhuangzi (c. 369-286 BCE) advised: "Resign yourself to what cannot be avoided and nourish what is within you – this is best. What more do you have to do to fulfill your mission? Nothing is as good as following orders (obeying fate) – that's how difficult it is!"

DONIA ZHANG,
March 2012

Appendix

Table A.1   The Dipper (seven stars and their two 'assistant stars')

English Name	Chinese Name	Chinese Astrological Name	Feng Shui Name	Yin Yang	Element	Characteristics
Dubhe	*tiānshū* 天枢	*pinlangxing* 贪狼星	*shengqi* 生气	yang	wood	extremely auspicious
Merak	*tiānxuán* 天璇	*jumenxing* 巨门星	*tianyi* 天医	yang	earth	moderate auspicious
Phecda	*tiānjī* 天玑	*lucunxing* 禄存星	*huohai* 祸害	yin	earth	moderate inauspicious
Megrez	*tiānquán* 天权	*wenquxing* 文曲星	*liusha* 六煞	yin	water	moderate inauspicious
Alioth	*yùhéng* 玉衡	*lianzhenxing* 廉贞星	*wugui* 五鬼	yin	fire	extremely inauspicious
Mizar	*kāiyáng* 开阳	*wuquxing* 武曲星	*yannian* 延年	yang	metal	moderate auspicious
Alkaid	*yáoguāng* 摇光	*pojunxing* 破军星	*jueming* 绝命	yin	metal	extremely inauspicious
Assistant star (left)	*zuofu* 左辅	*zuofu* 左辅	*zuofu* 左辅	yin	wood	little auspicious
Assistant star (right)	*youbi* 右弼	*youbi* 右弼	*youbi* 右弼	unsure	unsure	unsure

Note: for ease of use, the star names were translated in terms of their characteristics.

Table A.2 Natural elements and cultural connotations in Chinese house decorations

Natural Elements	Cultural Connotations
Apple and crabapple trees	Safety in every matter (事事平安 shi shi ping an)
Crabapple tree, the swastika 卐 (representing 万 wan, meaning 'ten thousand,' or bountiful and plentiful), and the characters 如意 (as you wish)	Everything according to your wish (万事如意 wan shi ru yi)
Peony and white-haired old man	Rich at old age (富贵白头 fu gui bai tou)
Glossy ganoderma, narcissus, bamboo, and peach-shaped birthday cake	Fairy congratulates (an elderly) on his/her birthday (灵仙祝寿 ling xian zhu shou)
Pine, bamboo, and plum tree	Three friends of winter (岁寒三友 sui han san you)
Plum, orchid, bamboo, and chrysanthemum	Four men of moral integrity (四君子 si jun zi)
Elephant and vase	Peace and tranquillity in presence (太平有象 tai ping you xiang)
Chrysanthemum and sparrow	Happy home (居家欢乐 ju jia huan le)
Gourd and vine	Thousands of descendants (子孙万代 zi sun wan dai)
Bat and pomegranate	Many offspring and good fortunes (多子多福 duo zi duo fu)

Natural Elements	Cultural Connotations
Vase and Chinese rose	Safety in all seasons (四季平安 si ji ping an)
Vase and quail	Safe and sound (平平安安 ping an an)
Quail, chrysanthemum, and Chinese sweet gum leaves	Live and work in peace and contentment (安居乐业 an ju le ye)
Peony and Chinese flowering crabapple	Wealth and rank fill the house (富贵满堂 fu gui man tang)
Gold and jade joined together	Gold and jade fill the hall (金玉同合 jin yu tong he, or 金玉满堂 jin yu man tang)
Pine and red-crowned crane	Pine and crane promise longevity (松鹤延年 he song he yan nian)
Pine, red-crowned crane, and deer (sika)	Crane and deer promise longevity (鹤鹿同春 he lu tong chun)
Plum blossom and magpie	Joy shows on the tip of the brow (喜上眉梢 xi shang mei shao)
Two spiders ('joy')	When joy arrives under the eaves, it is always double (喜到檐前每是双 xi dao yan qian mei shi shuang)
Horse, monkey, pine, bee, and seal ribbon	Awarding official rank and hanging seal (封侯挂印 feng hou gua yin)

Natural Elements	Cultural Connotations
Monkey riding a horse with bees flying on the top	Awarding official rank immediately (马上封侯 *ma shang feng hou*)
Cock (rooster) and cockscomb	Promotion of official position (官上加官 *guan shang jia guan*)
Longan, litchi, and walnut	Obtaining excellent scores successively for three times (连中三元 *lian zhong san yuan*)
Bat, deer, and peach	Good fortune, emolument, and longevity (福,禄,寿 *fu, lu, shou*)
Bat, the character 寿 (*shou*), and ribbon	Continuous good fortune and longevity (福寿绵长 *fu shou mian chang*)
Bats and the character 寿 (*shou*)	Five good fortunes holding longevity in their hands (五福捧寿 *wu fu peng shou*)
Character 寿 (*shou*) and the swastika 卍	Thousands of good fortunes and longevity (万福万寿 *wan fu wan shou*)

Note: This idiom derived from an ancient examinee that came first in the provincial imperial examination (解元), first in the regional imperial examination (会元), and first in the national imperial examination (状元).

Source: My summary and translations from Chavannes, 1973; Ma, 1999; Knapp, 1999, 2005a; Suzhou Housing Management Bureau, 2004

Table A.3   Chinese 24 solar seasons

Chinese Name	English Translation	Gregorian Date	Sun Longitude
立春	Start of Spring	February 3-5	315
雨水	Rain Water	February 18-20	330
惊蛰	Waking of Insects	March 5-7	345
春分	Spring Equinox	March 20-22	0
清明	Clear Bright	April 4-6	15
谷雨	Grain Rain	April 19-21	30
立夏	Start of Summer	May 5-7	45
小满	Grain Full	May 20-22	60
芒种	Grain in Ear	June 5-7	75
夏至	Summer Solstice	June 20-21	90
小暑	Slight Heat	July 6-8	105
大暑	Great Heat	July 22-24	120
立秋	Start of Autumn	August 7-9	135
处暑	Limit of Heat	August 22-24	150
白露	White Dew	September 7-9	165
秋分	Autumnal Equinox	September 22-24	180
寒露	Cold Dew	October 8-9	195
霜降	Frost Descent	October 23-24	210
立冬	Start of Winter	November 7-8	225
小雪	Slight Snow	November 22-23	240
大雪	Great Snow	December 6-8	255
冬至	Winter Solstice	December 21-22	270
小寒	Slight Cold	January 5-7	285
大寒	Great Cold	January 20-21	300

Source: Chinese Dictionary, 2006

Table A.4   Han Chinese cultural festivals on lunar and solar calendars

Seasonal Celebrations	Festival Name in Usual English Translation	Chinese Lunar Date	Gregorian Solar Date
Spring Celebrations	Spring Festival (Lunar New Year)	1st day of 1st lunar month	
	Lantern Festival (Spring Spirit Festival)	15th day of 1st lunar month	
	Blue Dragon Festival (Dragon Raising its Head)	2nd day of 2nd lunar month	Related to 'Waking of Insects' on March 5-7
	Qing Ming Festival (Clear Brightness Day)		April 5; April 4 on leap years
Summer Celebrations	Dragon Boat Festival	5th day of 5th lunar month	Related to Summer Solstice June 20-21
	Bathing and Basking Festival	6th day of 6th lunar month	
	Night of Sevens (Chinese Valentine's Day)	7th day of 7th lunar month	
	Summer Lantern Festival (Mid-Summer Spirit Festival)	15th day of 7th lunar month	

Seasonal Celebrations	Festival Name in Usual English Translation	Chinese Lunar Date	Gregorian Solar Date
Autumn Celebrations	Mid-Autumn Festival (Moon Festival)	15th day of 8th lunar month	
	Double Ninth Festival (Double Yang Festival)	9th day of 9th lunar month	
	Water Lantern Festival (Autumn Spirit Festival)	15th day of 10th lunar month	
Winter Celebrations	Winter Solstice Festival (Midwinter Festival)		December 21-22
	Laba Festival (Congee Festival)	8th day of 12th lunar month	
	Thanksgiving to the Stove God	23rd day of 12th lunar month	

References

Abercrombie, N., Hill, S., and Turner, B.S. (2000). *Penguin dictionary of sociology* (4th ed.). London: Penguin.

Abramson, D.B. (1997). *Neighborhood redevelopment as a cultural problem: a Western perspective on current plans for the old city of Beijing.* PhD thesis, Institute of Architectural and Urban Studies, Tsinghua University, Beijing, China.

Abramson, D.B. (2001). Beijing's preservation policy and the fate of the siheyuan. *Traditional Dwellings and Settlements Review: Journal of the International Association for the Study of Traditional Environments,* 13(1), pp. 7-22.

Abramson, D.B. (2007). The aesthetics of city-scale preservation policy in Beijing. *Planning Perspectives,* 22 (2), pp. 129-166.

Acharya, S.K. (2005). Urban development in post-reform China: insights from Beijing. *Norwegian Journal of Geography,* 59, pp. 228-236.

Across China (走遍中国, 2002) Episode 9: Zhejiang Ningbo 10-li red dress. TV series produced by CCTV (China Central Television).

Agenda 21 for Culture (2004). *An undertaking by cities and local governments for cultural development.* Retrieved April 17, 2007 from: http://www.agenda21cultura.net

Ahmad, D. and Zong, Y. (1995). *An overview of the construction industry in China.* Retrieved July 2, 2011 from: http://cibworld.xs4all.nl/dl/ib/9701/pages/31.htm

Al-Hagla, K.S. (2005). *Cultural sustainability: an asset of cultural tourism industry.* International Centre for Research on the Economics of Culture, Institutions, and Creativity (EBLA), Department of Economics, Università di Torino.

Al-Masri, W. (2010). The courtyard house in Kuwait today: design approaches and case studies. In N.O. Rabbat (Ed.), *The courtyard house: from cultural reference to universal relevance* (pp. 203-221). London: Ashgate.

Alexander, A., Hirako, Y., Dorje, L., and de Azevedo, P. (2003). *Beijing historic city study.* Asian Coalition for Housing Rights. Retrieved March 5, 2007 from: http://www.achr.net/UAC%20Web%20DLs/Beijing02a.doc

Alexander, C. Ishikawa, S., Silverstein, M., Jacobson, M., Fiksdahl-King, I., and Angel, S. (1977). *A pattern language.* New York: Oxford University Press.

Alred, J. (2008). *Sustainability charter for Powell River: base document.* Retrieved November 30, 2011 from: http://pr.viu.ca/communitysustainability/documents/Sustainability%20Charter%20for%20Powell%20River%20180808.pdf

Amato, P.R. (2008). *Recent changes in family structure: implications for children, adults, and society.* National Healthy Marriage Resource Center. Retrieved May

30, 2011 from: http://www.healthymarriageinfo.org/docs/changefamstructure. pdf

Amin, N., King, N., and Zhang, D. (1994). Investigating dimensions for robust blocks. In Amin *et al. The definition and achievement of good mixed-use town* (pp. 61-113). Issues Project, Joint Centre for Urban Design, Oxford Brookes University, UK.

Anderson, E.N. (2001). Flowering apricot: environmental practice, folk religion, and Daoism. In N.J. Girardot, J. Miller, and X. Liu (Eds.), *Daoism and ecology: ways within a cosmic landscape* (pp. 157-183). Cambridge, MA: Harvard University Press.

Architectural Design Code for Fire Protection (1987, 2006) (建筑设计防火规范). Retrieved August 2008 from: http://bbs.co.163.com/content/361_517373_1. html

Attmann, O. (2009). *Green architecture: advanced technologies and materials* (Mcgraw-Hill's Greensource). Ohio: McGraw-Hill Professional.

Bai, H. (2007). *Living in the old Beijing* (老北京的居住, Chinese edition). Beijing: Yanshan Publishing House.

Bai, X. (2005a). Pure Brightness festival. In H. Jin, Y. He, and X. Bai, *The traditional Chinese festivals and tales* (pp. 26-31). Sichuan: Chongqing Publishing House.

Bai, X. (2005b). Mid-Autumn festival. In H. Jin, Y. He, and X. Bai, *The traditional Chinese festivals and tales* (pp. 40-44). Sichuan: Chongqing Publishing House.

Bai, X. (2005c). Zhongyuan festival. In H. Jin, Y. He, and X. Bai, *The traditional Chinese festivals and tales* (pp. 53-57). Sichuan: Chongqing Publishing House.

Baker, H. (1979). *Chinese family and kinship*. London: MacMillian Press.

Barnard, A., and Spencer, J. (Eds.) (1997). *Encyclopedia of social and cultural anthropology*. London: Routledge.

Beijing City Planning and Construction Review (2004). Focus on Nanchizi (观照南池子). 北京规划建设 *Beijing City Planning and Construction Review*, 2, pp. 98-99.

Beijing City Planning Chart 1949-2005 (2007). Beijing Municipal Planning Commission, Beijing Municipal Urban Planning and Design Institute, Beijing City Planning Institute.

Beijing City Planning Committee (2002). *Conservation planning of 25 historic areas in Beijing's old city* (《北京旧城二十五片历史文化保护区保护规划》). Beijing: Yan Shan Publishing House.

Beijing Culture Net (2002). *Beijing view: some thoughts on the remaking of hutongs in the old city of Beijing* (北京视点: 对北京旧城—胡同改造的一点想法). March 16, 2002. Retrieved February 14, 2008 from: http://www.oldbj. com/bjview/ReadNews.asp?NewsID=336andBigClassName=andBigClassID =23andSmallClassID=30andSmallClassName=andSpecialID=17

Beijing Famous Historic and Cultural City Protection Regulation (2002) (北京历史文化名城保护规划). *Beijing Evening Newspaper*, September 19, 2002.

Bekerman, Z. and Kopelowitz, E. (Eds.) (2008). *Cultural education -cultural sustainability: minority, diaspora, indigenous, and ethno-religious groups in multicultural societies*. New York and London: Routledge.

Bell, S. (2003). *Measuring sustainability: learning by doing*. London: Earthscan Publications.

Bergman, D. (2011). *Sustainable design: a critical guide for architects and interior, lighting, and environmental designers*. NY: Princeton Architectural Press.

Berthrong, J. (1998). Motifs for a new Confucian ecological vision. In M.E. Tucker and J. Berthrong (Eds.), *Confucianism and ecology: the interrelation of heaven, earth, and humans* (pp. 237-263). Cambridge, MA: Harvard University Press.

Bickford, M. (2005). The symbolic seasonal round in house and palace: counting the auspicious nines in traditional China. In R.G. Knapp and K.-Y. Lo (Eds.), *House home family: living and being Chinese* (pp. 349-371). Honolulu: University of Hawai'i Press.

Blaser, W. (1985). *Atrium: five thousand years of open courtyards*. New York: Wepf and Co. AG, Basel.

Blaser, W. (1995). *Courtyard house in China: tradition and present* (2nd enlarged ed.). Basel/Boston/Berlin: Birkhäuser Verlag.

Bodeen, C. (2011). *Beijing's Confucius statue mysteriously removed*. Yahoo! News Apr 23, 2011. Retrieved November 6, 2011 from: http://news.yahoo.com/beijings-confucius-statue-mysteriously-removed-062453293.html

Bosselmann, P. *et al* (1984). *Sun, wind, and comfort: a study of open spaces and sidewalks in four downtown areas*. Berkeley, CA: University of California Press.

Bourdieu, P. (1973). *Cultural reproduction and social reproduction*. London: Tavistock.

Brand, D. (2005). Four pillars in practice at the City of Port Phillip. In Cultural Development Network, *The fourth pillar of sustainability: culture, engagement and sustainable communities* (pp. 76-81). Report from the Fourth Pillar Conference, Melbourne, November 2004.

Branigan, T. (2012). Chinese developers demolish home of revered architects: demolition of house where Liang Sicheng and his wife Lin Huiyin once worked has horrified heritage experts. *The Guardian*, January 30, 2012. Retrieved February 2, 2012 from: http://www.guardian.co.uk/world/2012/jan/30/chinese-developers-demolish-home-architect

Bray, D. (2005). *Social space and governance in urban China: the danwei system from origins to reform*. Palo Alto: Stanford University Press.

Bray, F. (1997). *Technology and gender: fabrics of power in late imperial China*. Berkeley: University of California Press.

Bray, F. (2005). The inner quarters: oppression or freedom? In R.G. Knapp and K.-Y. Lo (Eds.), *House home family: living and being Chinese* (pp. 259-279). Honolulu: University of Hawai'i Press.

Broudehoux, A.-M. (1994). *Neighborhood regeneration in Beijing: an overview of projects implemented in the inner city since 1990*. Master of Architecture thesis, School of Architecture, McGill University, Montreal, Canada.

Bruce, S., and Yearley, S. (2006). *The sage dictionary of sociology*. London: Sage.

Burns, J.M. (1978). *Leadership*. New York: Harper and Row.

Canadian Commission for UNESCO (n.d.). *UN decade of education for sustainable development* (2005-2014). Retrieved November 6, 2011 from: http://www.unesco.ca/en/interdisciplinary/ESD/default.aspx

Cao, Y. (2001). *Let the ancient capital's style and features reveal its original splendor: report of the Beijing famous historic and cultural city protection* (让古都风貌重现辉煌——北京历史文化名城保护纪实). 人民教育出版社 新华社 People's Education Publisher, Xinhua News, 2001-02-21.

Capra, F. (1975/1999). *Tao of physics: an exploration of the parallels between modern physics and Eastern mysticism*. Boston, MA: Shambhala Publications.

Carmona, M., Heath, T., Oc, T., and Tiesdell, S. (2003). *Public places, urban spaces: the dimensions of urban design*. Oxford: Architectural Press.

Carroon, J. and Moe, R. (2010). *Sustainable preservation: greening existing buildings* (Wiley Books on Sustainable Design). NJ: Wiley.

Casault, A. (1987). The Beijing courtyard house: its traditional form and present state. *Open House International,* 12(1), pp. 31-41.

Casault, A. (1988). *Understanding the changes and constants of the courtyard house neighbourhoods in Beijing*. Master's thesis, MIT, Massachusetts, USA.

CCTV (2006). *The rise of the great nations* (大国崛起). TV documentary series by China Central Television.

Chan, C.-S. and Xiong, Y. (2007). The features and forces that define, maintain, and endanger Beijing courtyard housing. *Journal of Architectural and Planning Research,* 24(1), pp. 42-64.

Chan, W.-T. (Tr.) (1969). *A source book in Chinese philosophy*. NJ: Princeton University Press.

Chang, K.-Y. (1991). Confucianism in the Republic of China and its role in mainland China's reform. In S. Krieger and R. Trauzettel (Eds.), *Confucianism and the modernization of China* (pp. 229-242). Mainz, Germany: Hase and Koehler Verlag.

Chang, S.S.-H. (1986). *Spatial organisation and socio-cultural basis of traditional courtyard houses*. PhD thesis, University of Edinburgh, UK.

Characteristics of Timber-Structured Chinese Ancient Buildings (中国古建木构 特征概说, 2008-09-05). Retrieved September 10, 2010 from: http://hi.baidu.com/chidori69/blog/item/5dc4cc3f1450f0ea55e7234f.html

Chatfield-Taylor, A. (1981). Vernacular architecture and historic preservation in modern China. *Ekistics,* 48(288), pp. 199-201.

Chavannes, E. (1973). *The five happinesses: symbolism in Chinese popular art* (Trans. E.S. Atwood). New York: Weatherhill.

Chen, F. (2011). Traditional architectural forms in market oriented Chinese cities: place for localities or symbol of culture? *Habitat International*, 35(2), pp. 410-418.

Chen, F. and Romice, O. (2009). Preserving the cultural identity of Chinese cities in urban design through a typomorphological approach. *Urban Design International*, 14(1), pp. 36-54.

Cheng, C.-Y. (1998). The trinity of cosmology, ecology, and ethics in the Confucian personhood. In M.E. Tucker and J. Berthrong (Eds.), *Confucianism and ecology: the interrelation of heaven, earth, and humans* (pp. 211-235). Cambridge, MA: Harvard University Press.

China Classical Tours (2006). *Suzhou*. Retrieved February 28, 2007 from http://www.china-travel.cc/City_guide/Suzhou/index.htm

China Daily (2010-04-12). *Poor construction quality keeps foreign property buyers away*. Retrieved July 2, 2011 from: http://www.chinadaily.com.cn/metro/2010-04/12/content_9715734.htm

China Daily (2010-08-07). *'Most homes' to be demolished in 20 years*. Retrieved July 2, 2011 from: http://www.chinadaily.com.cn/china/2010-08/07/content_11113982.htm

China.org.cn (n.d.). *Beijing 2005: the year in review*. Retrieved December 30, 2011 from: http://www.china.org.cn/english/features/ProvinceView/163623.htm

China Tourist Cities (2002). *Suzhou*. Retrieved February 28, 2007 from http://www.asia-planet.net/china/suzhou.htm

ChinaTravel.com (n.d.). *Beijing Overview*. Retrieved January 12, 2009 from: http://www.chinatravel.com/beijing/overview.htm

ChinaTravel.com (n.d.). *Suzhou Overview*. Retrieved January 12, 2009 from: http://www.chinatravel.com/jiangsu/suzhou/

Chinese Academy of Social Sciences (2003). *Science and culture of the 20th century China*. Retrieved November 3, 2011 from: http://bic.cass.cn/english/InfoShow/Arcitle_Show_Forum2_Show.asp?ID=264&Title=The%20Humanities%20Study&strNavigation=Home-%3EForum-%3ECultrue%20and%20Art&BigClassID=4&SmallClassID=8

Chinese Historical and Cultural Project (2009). *Chinese wedding traditions*. Retrieved April 1, 2009 from: http://www.chcp.org/wedding.html

Ching, J. (1993). *Chinese religions*. London: Macmillan.

Ch'ü, H.-Y. (1994). *Report on the fifth implementation of the second section of the plan for a basic survey on social change in Taiwan*. Taipei: Chung-yang yen-chiu min-tsu-hsüeh yen-chiu-so.

Civil Design General (1987, 2005) (民用建筑设计通则). Retrieved August 2008 from: http://www.law110.com/lawserve/guihua/1800056.htm and http://ziliao.hopebook.net/ziliao-10800.htm

Cohen, M.L. (1990). Lineage organization in north China. *Journal of Asian Studies*, 49(3), pp. 509-534.

Cohen, M.L. (1992). Family management and family division in contemporary rural China. *China Quarterly*, 130, pp. 357-377.

Cohen, M.L. (1998). North China rural families: changes during the communist era. *Études Chinoises*, 17(1-2), pp. 60-154.

Cohen, M.L. (2005). House united, house divided: myths and realities, then and now. In R.G. Knapp and K.-Y. Lo (Eds.), *House home family: living and being Chinese* (pp. 235-257). Honolulu: University of Hawai'i Press.

Collins, M. (2005). Protecting the ancient alleys of Beijing. *Contemporary Review*, 286(1668), pp. 34-38.

Confucius (551-479 BCE). *Confucian analects, the great learning, and the doctrine of the mean* (translated and annotated by J. Legge, 1893/1971). New York: Dover Publications.

Confucius Institute Online (2009). Retrieved November 3, 2011 from: http://www.chinese.cn/

Creative City Network of Canada (2004a). *Creating economic and social benefits for communities*. Special Edition 1.

Creative City Network of Canada (2004b). *Urban revitalization and renewal*. Special Edition 1.

Creative City Network of Canada (2004c). *Quality of place, quality of life*. Special Edition 1.

Creative City Network of Canada (2005a). *Nurturing culture and creativity to build community*. Special Edition 2.

Creative City Network of Canada (2005b). *Integrate culture into the planning of cities and communities*. Special Edition 2.

Creative City Network of Canada (2006a). *Culture: transforming lives, sustaining communities*. Special Edition 3.

Creative City Network of Canada (2006b). *Culture: the forth pillar of sustainability*. Special Edition 3.

Creative City Network of Canada (2007a). *Models of sustainability incorporating culture*. Special Edition 4.

Creative City Network of Canada (2007b). *Key contexts*. Special Edition 4.

Creswell, J.W. (2002). *Educational research: planning, conducting, and evaluating quantitative and qualitative research*. Upper Saddle River, NJ: Pearson Education.

Csongor, B. (1991). On the limits of anti-Confucianism. In S. Krieger and R. Trauzettel (Eds.), *Confucianism and the modernization of China* (pp. 449-453). Mainz, Germany: Hase and Koehler Verlag.

Cuisenier, J. (1997). Spatial. In P. Oliver (Ed.), *Encyclopedia of vernacular architecture of the world: theories and principles* (Vol. 1) (pp. 60-62). Cambridge: Cambridge University Press.

Cultural Development Network (2005). *The fourth pillar of sustainability: culture, engagement and sustainable communities*. Report from the Fourth Pillar Conference, Melbourne, November 2004. Retrieved September 27, 2011 from: http://www.culturaldevelopment.net.au/downloads/FPS_ConferenceRpt.pdf

Darlow, A. (1996). Cultural policy and urban sustainability: making a missing link? *Planning Practice and Research*, 11(3), 291-301.

Davey, P. (2000). Courtly life. *Architectural Review*, 207(1236), pp. 73-75.

Diamond, J. (1999). *Guns, germs, and steel: the fates of human societies*. New York: WW Norton.

Dines, N., Cattell, V., Gesler, W., and Curtis, S. (2006). *Public spaces, social relations and well-being in east London*. London: UK. Joseph Rowntree Foundation Report no. 1925.

Ding, S. (Ed.) (1992). *Chinese local annals: collected materials on customs, East China volume* (中国地方志民俗资料汇编, 华东卷). Beijing: Booklist and Documents Publishing House (书目文献出版社).

Dong, G. (1987). Beijing: housing and community development. *Ekistics*, 54(322), pp. 34-39.

du Cros *et al.* (2005). Cultural heritage assets in China as sustainable tourism products: case studies of the hutongs and the Huanghua section of the Great Wall. *Journal of Sustainable Tourism*, 13(2), pp. 171-254.

Du, Y. (2008). *Beijing investing large money renovating old siheyuan, 1/3 residents will move out of the old city* (北京投巨资做旧四合院, 1/3居民将迁出旧城区). 中新社 China News, November 19, 2008. Retrieved November 19, 2008 from: http://news.wenxuecity.com/messages/200811/news-gb2312-740968. html

Duxbury, N. (2003). *Cultural indicators and benchmarks in community indicator projects: performance measures for cultural investment?* Paper from the Accounting for Culture Colloquium, November 13-15, 2003 Retrieved September 29, 2011 from: http://www.haliburtoncooperative.on.ca/literature/pdf/S-663_Cultural_Indicators_andBenchmarks.pdf

Duxbury, N. and Gillette, E. (2007). *Culture as a key dimension of sustainability: exploring concepts, themes, and models*. Working paper published by the Creative City Network of Canada – Centre of Expertise on Culture and Communities.

Duxbury, N. and Jeannotte, M.S. (2011). Introduction: culture and sustainable communities. *Culture and Local Governance / Culture et gouvernance locale*, 3(1-2), pp. 1-10.

Duxbury, N., Simons, D., and Warfield, K. (2006). Local policies and expressions of cultural diversity: Canada and the United States. In UNESCO's report (September 20, 2006), *Local policies for cultural diversity* (pp. 32-51). UNESCO, UCLG and Barcelona City Council.

Earth Pledge (2001). *Sustainable architecture white papers: essays on design and building for a sustainable future* (Earth Pledge Foundation Series on Sustainable Development). NY: Earth Pledge Foundation.

East St. Louis Action Research Project (n.d.). *What is sustainable architecture?* Retrieved May 29, 2011 from: http://www.eslarp.uiuc.edu/arch/ARCH371-F99/groups/k/susarch.html

Ebrey, P.B. (1991). *Chu Hsi's Family Rituals: a 12th-century Chinese manual for the performance of cappings, weddings, funerals, and ancestral rites.* Princeton: Princeton University Press.

Economist (2009-10-23). *Motoring ahead: more cars are now sold in China than in America.* Retrieved March 10, 2011 from: http://www.economist.com/node/14732026?story_id=14732026&fsrc=nwl

Edwards, B., Sibley, M., Hakmi, M., and Land, P. (Eds.) (2006). *Courtyard housing: past, present, and future.* New York: Taylor and Francis.

Ekblad, S. and Werne, F. (1990a). Housing and health in Beijing: implications of high-rise housing on children and the aged. *Journal of Sociology and Social Welfare,* 17(1), pp. 51-77.

Ekblad, S. and Werne, F. (1990b). Housing in Beijing. *Ekistics,* 342/343, pp. 146-159.

Ekblad, S., Chen, C-H., Huang, Y-Q, and Li, S. (1992). Effects of dwelling types and facilities on crowding, stress, life satisfaction and health of households in western Beijing. *Ekistics,* 354/355, pp. 195-205.

Eliade, M. (1959). *The sacred and the profane: the nature of religion* (translated from the French by Willard R. Trask). Toronto: Harcourt Harvest Book.

Entwurf, G. (2005). *Gulou, Beijing.* Major design for the Central Academy of Fine Arts, Beijing.

Ettouney, S.M. (1975). *Courtyard acoustics and aerodynamics: an investigation of the acoustic and wind environments of courtyard housing.* PhD thesis, University of Sheffield, UK.

EurActiv (2004). *Sustainable development: introduction.* Retrieved March 3, 2010 from: http://www.euractiv.com/en/sustainability/sustainable-development-introduction/article-117539

Explore (June, 2007). *Garden villas* (园林别墅). 发现 (苏州版, pp. 12-13).

Explore (August, 2007). *Humanity manifests great fascination; details construct top-quality products: interview with Zhang Han* (人文彰显魅力，细节构筑精品：访苏州嘉德房地产开发有限公司副总经理张涵). 发现 (苏州版, pp. 20-22).

Explore (September, 2007). *Mid-autumn day and Suzhou people.* 发现 (苏州版, pp. 78-79).

Explore (October, 2007). *The long history of brick carving's gate tower* [sic]. 发现 (苏州版, pp. 82-89).

Fang, Y. (2004). *Residential satisfaction conceptual framework revisited: a study on redeveloped neighborhoods in inner city Beijing.* Toronto Conference Paper.

Fang, Y. (2006). Residential satisfaction, moving intention and moving behaviours: a study of redeveloped neighbourhoods in inner-city Beijing. *Housing Studies,* 21(5), pp. 671-694.

Faure, D. (2005). Between house and home: the family in south China. In R.G. Knapp and K.-Y. Lo (Eds.), *House home family: living and being Chinese* (pp. 281-293). Honolulu: University of Hawai'i Press.

Fei, X. (1947/1992). *From the soil: the foundations of Chinese society* (Tr. G. Hamilton and Z. Wang). Berkeley: University of California Press.

Field, S.L. (2001). In search of dragons: the folk ecology of fengshui. In N.J. Girardot, J. Miller, and X. Liu (Eds.), *Daoism and ecology: ways within a cosmic landscape* (pp. 185-200). Cambridge, MA: Harvard University Press.

Flath, J.A. (2005). Reading the text of the home: domestic ritual configuration through print. In R.G. Knapp and K.-Y. Lo (Eds.), *House home family: living and being Chinese* (pp. 325-347). Honolulu: University of Hawai'i Press.

Florida, R. (2002). *The rise of the creative class: and how it's transforming work, leisure, community and everyday life*. New York: Basic Books.

Florida, R. (2008). *Who's your city? How the creative economy is making where you live the most important decision of your life*. New York: Basic Books.

Flyvbjerg, B. (2006). Five misunderstandings about case study research. *Qualitative Inquiry*, 12(2), pp. 219-245.

Flyvbjerg, B. (2011). Case study. In N.K. Denzin and Y.S. Lincoln (Eds.), *The sage handbook of qualitative research* (4th ed., pp. 301-316). Thousand Oaks, CA: Sage.

Ford, B., Lau, B., and Zhang, H. (2006). *The environmental performance of traditional courtyard housing in China: case study Zhang's house, Zhouzhuang, Jiangsu Province*. PLEA2006 - The 23rd Conference on Passive and Low Energy Architecture, Geneva, Switzerland, 6-8 September 2006.

Fryer, S. (2007). *Briefing: how to achieve true quality in construction*. Proceedings of the Institution of Civil Engineers, Management, Procurement and Law 160. MPI pp. 5-6. London: Institution of Civil Engineers by Thomas Telford. Retrieved March 20, 2011 from: http://www.epdconstruction.com/articles/quality.pdf

Fu, X., Guo, D., Liu, X., Pan, G., Qiao, Y., and Sun, D. (2002). *Chinese architecture* (Translated and Edited by N.S. Steinhardt). New Haven: Yale University Press.

Fu, X. (2002a). The three kingdoms, western and eastern Jin, and northern and southern dynasties. In X. Fu, *et al.*, *Chinese architecture* (pp. 61-89). New Haven: Yale University Press.

Fu, X. (2002b). The Sui, Tang, and five dynasties. In X. Fu, *et al.*, *Chinese architecture* (pp. 91-133). New Haven: Yale University Press.

Gao, M. (1999). Manufacturing of truth and culture of the elite. *Journal of Contemporary Asia*, 29(3), pp. 309-328.

Gates, H. (1996). *China's motor: a thousand years of petty capitalism*. Ithaca, NY: Cornell University Press.

Gaubatz, P. (1995). Changing Beijing. *Geographical Review*, 85(1), pp. 79-97.

Gaubatz, P. (1999). China's urban transformation: patterns and processes of morphological change in Beijing, Shanghai and Guangzhou. *Urban Studies*, 36(9), pp. 1495-1521.

Geertz, C. (1973). *Interpretation of cultures*. New York: Basic Books.

Gehl, J. (1971/2001). *Life between buildings: using public space* (translated by J. Koch). Copenhagen, Denmark: Arkitektens Forlag (The Danish Architectural Press).

Geisler, T.C. (2000). On public toilets in Beijing. *Journal of Architectural Education,* 53(4), pp. 216-219.

Gharai, F. (1998). *The value of neighbourhoods: a cultural approach to urban design.* Department of Architectural Studies, University of Sheffield, UK.

Giedion, S. (1981). *The beginnings of architecture.* Bollingen Series XXX. 6.11, Cambridge, MA: Harvard University Press.

Girardot, N.J., Miller, J. and Liu, X. (Eds.) (2001). *Daoism and ecology: ways within a cosmic landscape.* Cambridge, MA: Harvard University Press.

Gissen, D. (2003). *Big and green: toward sustainable architecture in the 21st century.* NY: Princeton Architectural Press.

Glassie, H. (1997). Aesthetic. In In P. Oliver (Ed.), *Encyclopedia of vernacular architecture of the world: theories and principles* (Vol. 1) (pp. 3-5). Cambridge: Cambridge University Press.

Goh, P.K. (comp.) (2004). *Origins of Chinese festivals* (translated by K.K. Koh; illustrated by C. Fu). Singapore: Asiapac Books.

Gong, D. (1991). Confucius' humanitarianist ideas and the contemporary international community. In S. Krieger and R. Trauzettel (Eds.), *Confucianism and the modernization of China* (pp. 314-323). Mainz, Germany: Hase and Koehler Verlag.

Gong, Q. (1999). Detailed planning of Suzhou Tongfang Xiang residential district regeneration (苏州古城桐芳巷居住街坊改造详细规划). In Urban Planning Research Centre (Ed.), *Collection of outstanding urban planning cases* (城市规划精品集锦). Beijing: China Architectural Industry Press.

Groat, L. and Wang, D. (2002). *Architectural research methods.* NJ: Wiley.

Gu, C. and Liu, H. (2002). Social polarization and segregation in Beijing. In J.R. Logan (Ed.), *The new Chinese city: globalization and market reform* (pp. 198-211). Oxford: Blackwell Publishers.

Gudykunst, W. and Kim, Y. (1984). *Communicating with strangers: an approach to intercultural communication.* Reading, MA: Addison-Wesley.

Guo, D. (2002). The Liao, Song, Xi Xia, and Jin dynasties. In X. Fu, *et al., Chinese architecture* (pp. 135-197). New Haven: Yale University Press.

Guo, Y. (2006). The next great leap forward: rapid growth means big challenges for China's transportation system. *Planning,* May 2006, pp. 22-27.

Hall, E.T. (1959). *The silent language.* Garden City, NY: Doubleday and Company.

Hall, E.T. (1966). *The hidden dimension.* Garden City, NY: Doubleday and Company.

Han, B. (2001). *Reweaving the fabric: a theoretical framework for the study of the social and spatial networks in the traditional neighbourhoods in Beijing, China.* Georgia Institute of Technology, USA.

Hargreaves, A. and Webster, R. (2000). *Social sustainability and local distinctiveness: arguments for the positive evaluation of place-centered awareness*. Paper presented at the ENHR 2000 Conference, 26-30 June 2000, Gavle.

Harrison, J. (2005). Construction quality assurance: white paper. *Performance Validation*. Retrieved March 20, 2011 from: http://www.perfval.com/news/ ConstructionQualityAssurance_WhitePaper_2005.pdf

Hartig, T. and Marcus, C.C. (2006). Healing gardens: places for nature in health care. *Lancet, Supplement, 68*, pp. 36-37.

Hawkes, J. (2001). *The fourth pillar of sustainability: culture's essential role in public planning*. Australia: Common Ground.

Hayward, R.S. (1994). *Housing typology: a rule of thumb for designers* (pamphlet). Joint Centre for Urban Design, Oxford Brookes University.

He, H. (1990). Living conditions in Beijing's courtyard housing. *Building in China, 3*(4), pp.14-20, 24.

He, H. (1993). Social and economic aspects of Beijing old-city redevelopment programs. *Building in China, 6*(3-4), pp. 16-23.

He, Y. (2005a). Lantern festival (translated by H. Jin). In H. Jin, Y. He, and X. Bai, *The traditional Chinese festivals and tales* (pp. 18-25). Sichuan: Chongqing Publishing House.

He, Y. (2005b). Dragon Boat festival (translated by H. Jin). In H. Jin, Y. He, and X. Bai, *The traditional Chinese festivals and tales* (pp. 32-39). Sichuan: Chongqing Publishing House.

He, Y. (2005c). Double Ninth day (translated by H. Jin). In H. Jin, Y. He, and X. Bai, *The traditional Chinese festivals and tales* (pp. 45-52). Sichuan: Chongqing Publishing House.

Heidari, S. (2000). *Thermal comfort in Iranian courtyard housing*. PhD thesis, University of Sheffield, UK.

Hester, R.T. (2006). *Design for ecological democracy*. Cambridge, MA: The MIT Press.

Higgs, E. (2003). *Nature by design: people, natural process, and ecological restoration*. Cambridge, MA: The MIT Press.

History of Chinese Architecture (1986/2009) by History of Chinese Architecture editorial group. Beijing: China Architecture and Building Press.

Ho, P.-P. (2003). China's vernacular architecture. In R.G. Knapp (Ed.), *Asia's old dwellings: tradition, resilience, change* (pp. 319-346). New York: Oxford University Press.

Ho, P.-P. (2005). Ancestral halls: family, lineage, and ritual. In R.G. Knapp and K.-Y. Lo (Eds.), *House home family: living and being Chinese* (pp. 295-323). Honolulu: University of Hawai'i Press.

Holt, R. (2005). The fourth pillar in three countries: Australia: the Australian context. In Cultural Development Network, *The fourth pillar of sustainability: culture, engagement and sustainable communities* (pp. 38-41). Report from the Fourth Pillar Conference, Melbourne, November 2004.

Hou, R. (2006). Stories of old houses. *China Today, 55* (4), pp. 20-23.

Hough, M. (1995/2004). *Cities and natural process* (2nd ed.). London and New York: Routledge.

Hsin, C.S.M. (2004). *Courtyard house with large south facing glazing: an ideal passive solar design of housing for thermal comfort: Milton Keynes energy world revisited.* M.Sc. thesis, Built Environment: Environmental Design and Engineering, University College London, UK.

Huang, W., Shang, K., Nan, W., and Pan, J. (1992). *China vernacular dwellings: Fujian and Guangdong provinces* (闽粤民宅). China: Tianjin Science and Technology Press.

Huntington, S.P. (1996/2011). *The clash of civilizations and the remaking of world order.* New York, NY: Simon and Schuster.

I Ching [*Yi Jing*] (1882) (Translated by J. Legge). Oxford: Clarendon Press.

I Ching or *Book of Changes* (1977) (Translated by R. Wilhelm and C.F. Baynes). New Jersey: Princeton University Press.

ICOMOS (1987). *Charter for the conservation of historic towns and urban areas* (Washington Charter 1987). Paris: ICOMOS.

ICOMOS (1999). *Charter on the built vernacular heritage.* Paris: ICOMOS.

Illa, S.M. and Ulldemolins, J.R. (2011). Cultural planning and community sustainability: the case of the cultural facilities plan of Catalonia. *Culture and Local Governance / Culture et gouvernance locale*, 3(1-2), pp. 71-82.

International Institute for Sustainable Development (2009). *What is sustainable development?* Retrieved March 3, 2010 from: http://www.iisd.org/sd/

ISCO (1988). *International standard classification of occupations.* International Labour Office.

ISCO (2008). *Updating the International Standard Classification of Occupations.* International Labour Office.

Jackson, P. (1989/1992). *Maps of meaning: an introduction to cultural geography.* London: Routledge.

Jacobs, J. (2004). *Dark age ahead.* Toronto: Random House.

Jary, D., and Jary, J. (Eds.) (2005). *Collins dictionary of sociology* (3rd ed.) Glasgow: HarperCollins.

Jervis, N. (1992). Dacaiyuan village, Henan: migration and village renewal. In R.G. Knapp (Ed.), *Chinese landscapes: the village as place* (pp. 245-257). Honolulu: University of Hawai'i Press.

Jervis, N. (2005). The meaning of *jia*: an introduction. In R.G. Knapp and K.-Y. Lo (Eds.), *House home family: living and being Chinese* (pp. 223-233). Honolulu: University of Hawai'i Press.

Jin, H. (2005). Spring festival. In H. Jin, Y. He, and X. Bai, *The traditional Chinese festivals and tales* (pp. 1-17). Sichuan: Chongqing Publishing House.

Jin, L., Feng, C., Chen, H., Dong, Q., Li, D. Qian, Y., and Hu, Y. (2004). *Jiangnan survey report.* Beijing, China: Tsinghua University, Institute of Architectural and Urban Studies, The 'Oriental Spirit' Research Group.

Johnson, A.G. (2000). *The Blackwell dictionary of sociology: a user's guide to sociological language* (2nd ed.). Malden, MA: Blackwell.

Johnston, R.J., Gregory, D., Pratt, G., and Watts, M. (Eds.) (2000). *The dictionary of human geography* (4th ed.). Oxford: Blackwell.

Johnston, D. and Gibson, S. (2008). *Green from the ground up: sustainable, healthy, and energy-efficient home construction* (Builder's Guide). CT: Taunton.

Kadekodi, G.K. (1992). *Paradigms of sustainable development.* Journal of SID, 3 pp. 72-76.

Kanazawa, S. and Che, J. (2002). Comparative study on residents' perception and activities in their outdoor spaces: cases of traditional blocks and a new housing project in Beijing. *Journal of Asian Architecture and Building Engineering,* 1(1), pp. 221-228.

Kates, R., Parris, T. and Leiserowitz, A. (2005). *What is sustainable development?* Environment 47(3), pp. 8-21. Retrieved March 3, 2010 from: http://www.hks. harvard.edu/sustsci/ists/docs/whatisSD_env_kates_0504.pdf

Keswick, M. (2003). *Chinese garden: history, art and architecture.* Cambridge, MA: Harvard University Press.

Kibert, C.J. (2011). *Sustainable construction: green building design and delivery* (book and WileyCPE.com course bundle). NJ: Wiley.

Kim, S.-Y. (2001). *Optimising courtyard housing design for solar radiation within dense urban environments.* PhD thesis, University of Sheffield, UK.

King, A.D. (2004). *Spaces of global cultures: architecture, urbanism, identity.* London: Spon.

Kingma, O. (2005). Restructuring communities: policies for a different society. In Cultural Development Network, *The fourth pillar of sustainability: culture, engagement and sustainable communities* (pp. 60-65). Report from the Fourth Pillar Conference, Melbourne, November 2004.

Khattab, O. (2002). Reconstruction of traditional architecture: a design education tool. *Global Built Environment Review,* 2(2), pp. 29-39.

Knapp, R.G. (1986). *China's traditional rural architecture: a cultural geography of the common house.* Honolulu: University of Hawai'i Press.

Knapp, R.G. (1999). *China's living houses: folk beliefs, symbols, and household ornamentation.* Honolulu: University of Hawai'i Press.

Knapp, R.G. (2000). *China's old dwellings.* Honolulu: University of Hawai'i Press.

Knapp, R.G. (Ed.) (2003). *Asia's old dwellings: tradition, resilience, change.* New York: Oxford University Press.

Knapp, R.G. (2005a). *Chinese houses.* North Clarendon, VT: Tuttle Publishing.

Knapp, R.G. (2005b). China's houses, homes, and families. In R.G. Knapp and K.-Y. Lo (Eds.), *House home family: living and being Chinese* (pp. 1-9). Honolulu: University of Hawai'i Press.

Knapp, R.G. (2005c). In search of the elusive Chinese house. In R.G. Knapp and K.-Y. Lo (Eds.), *House home family: living and being Chinese* (pp. 37-71). Honolulu: University of Hawai'i Press.

Knapp, R.G. (2005d). Siting and situating a dwelling: fengshui, house-building rituals, and amulets. In R.G. Knapp and K.-Y. Lo (Eds.), *House home family: living and being Chinese* (pp. 99-137). Honolulu: University of Hawai'i Press.

Knowles, R.L. (1998). *Rhythm and ritual: a motive for design.* Retrieved May 12, 2010 from: http://www-rcf.usc.edu/~rknowles/rhythm_ritual/rhy_rit.html

Knowles, R.L. (1999). *Rituals of place.* Retrieved May 12, 2010 from: http://www-rcf.usc.edu/~rknowles/rituals_place/rituals_place.html

Ko, D. (1992). Pursuing talent and virtue: education and women's culture in 17th- and 18th-century China. *Late Imperial China,* 13(1), pp. 9-39.

Ko, D. (1994). *Teachers of the inner chambers: women and culture in 17th-century China.* Stanford: Stanford University Press.

Kohn, L. (Comp.) (2001). Change starts small: Daoist practice and the ecology of individual lives. A roundtable discussion with Liu Ming, René Navarro, Linda Varone, Vincent Chu, Daniel Seitz, and Weidong Lu. In N.J. Girardot, J. Miller, and X. Liu (Eds.), *Daoism and ecology: ways within a cosmic landscape* (pp. 373-390). Cambridge, MA: Harvard University.

Kohn, L. (2008). *Laughing at the Dao: debates among Buddhists and Daoists in Medieval China.* Magdalena, NM: Three Pines Press.

Kong, F. (2004). *Dilapidated housing renewal and the conservation of traditional courtyard houses* (危房改造方式与传统四合院保护). 北京市文物局 Beijing Municipal Administration of Cultural Heritage. Retrieved February 13, 2008 from http://www.bjww.gov.cn/2004/12-7/3717-2.shtml

Körner, B. (1959). *Die religiöse welt der baüerin in Nordchina. Reports from the Scientific Expedition to the north-western provinces of China under the leadership of Sven Hedin, 8: Ethnography.* Stockholm: State Ethnographic Museum.

Kou, X. (2005). *A treasure dictionary for prosperous residences: a guide to residential Feng Shui* (旺宅宝典：住宅风水指南, Chinese edition). Beijing: Culture and Art Press.

Krieger, S. and Trauzettel, R. (Eds.) (1991). *Confucianism and the modernization of China.* Mainz, Germany: Hase and Koehler Verlag.

Kuang, Y. (1991). Modern values of the positive elements in Confucius' ideas concerning the study of man. In S. Krieger and R. Trauzettel (Eds.), *Confucianism and the modernization of China* (pp. 7-17). Mainz, Germany: Hase and Koehler Verlag.

Kubin, W. (1991). On the problem of the self in Confucianism. In S. Krieger and R. Trauzettel (Eds.), *Confucianism and the modernization of China* (pp. 63-95). Mainz, Germany: Hase and Koehler Verlag.

Kuwako, T. (1998). The philosophy of environmental correlation. In M.E. Tucker and J. Berthrong (Eds.), *Confucianism and ecology: the interrelation of heaven, earth, and humans* (pp. 151-168). Cambridge, MA: Harvard University Press.

Kwok, A.G. and Grondzik, W.T. (2007). *The green studio handbook: environmental strategies for schematic design.* Burlington, MA: Architectural Press.

Laaksonen, A. (2006). Local policies for cultural diversity: with emphasis on Latin America and Europe. In UNESCO's report (September 20, 2006), *Local policies for cultural diversity* (pp. 52-71). UNESCO, UCLG and Barcelona City Council.

Lai, C.T. (2001). The Daoist concept of central harmony in the *Scripture of Great Peace*: human responsibility for the maladies of nature. In N.J. Girardot, J. Miller, and X. Liu (Eds.), *Daoism and ecology: ways within a cosmic landscape* (pp. 95-111). Cambridge, MA: Harvard University.

Laozi (c.571-471 BCE). *Tao te ching: an illustrated journey* (translated by S. Mitchell, 1999). New York: HarperCollins.

Land, P. (2006). Courtyard housing: an 'afterthought.' In B. Edwards *et al.* (Eds.). *Courtyard housing: past, present, and future*. New York: Taylor and Francis.

Landry, C. (2000). *The creative city: a toolkit for urban innovators*. London: Earthscan.

Laroche, Y. (2005). The fourth pillar in three countries: Canada: adapted from the keynote address to the Creative City Network, Canada. In Cultural Development Network, *The fourth pillar of sustainability: culture, engagement and sustainable communities* (pp. 42-45). Report from the Fourth Pillar Conference, Melbourne, November 2004.

Lau, K.-K. (1991). An interpretation of Confucian virtues and their relevance to China's modernization. In S. Krieger and R. Trauzettel (Eds.), *Confucianism and the modernization of China* (pp. 210-228). Mainz, Germany: Hase and Koehler Verlag.

Law Bridge (1997-11-1). *Construction law of the People's Republic of China*. Retrieved July 2, 2011 from: http://www.law-bridge.net/english/LAW/20064/022246219215.html

Lechner, N. (2008). *Heating, cooling, lighting: sustainable design methods for architects*. NJ: Wiley.

Lee, C.Y. (2004). Foreword. In P.K. Goh (compiled), *Origins of Chinese festivals* (translated by K.K. Koh; illustrated by C. Fu). Singapore: Asiapac Books.

Legge, J. (Tr.) (1960). *The Chinese classics* (vol. 3). Hong Kong: Hong Kong University Press.

LeVine, R.A. (1984). Properties of culture: an ethnographic view. In R. Schwader and R. Levine (eds), *Culture theory: essays on mind, theory, and emotion* (pp. 67-87). Cambridge: Cambridge University Press.

Li, S.-P. (1993). Ch'ing cosmology and popular precepts. In R.J. Smith and D.W.Y. Kwok (Eds.), *Cosmology, ontology and human efficacy: essays in Chinese thought* (pp. 113-139). Honolulu: University of Hawai'i Press.

Li, T. (1998). *Residential renewal in old Chinese cities since 1979: under the transition from central-planned to market-driven economy*. Master of Urban and Regional Planning thesis, Virginia Polytechnic Institute and State University, Blacksburg, Virginia, USA.

Li, Z. (2007). *Report on eastern district historical and cultural style and features protection and dilapidated housing regeneration* (关于东城区历史文化风貌

保护和危旧房改造建设情况的调研报告）. Retrieved on February 13, 2008 from: http://www.bjdch.gov.cn/n4279063/n4279168/n4279801/4409955.html

Li, Z. and Wu, F. (2006). Socio-spatial differentiation and residential inequalities in Shanghai: a case study of three neighbourhoods. *Housing Studies,* 21(5), pp. 695-717.

Liang, J., and Zong, S. (2005). Recent conditions of Juer Hutong (菊儿胡同近况). 《北京规划建设》 *Beijing Planning Review*, 4, pp. 71-73.

Liang, S.C. (1998). *Frozen music* (凝动的音乐, Chinese edition). Tianjin: Hundred Flowers Literature and Art Publishing House.

Lin, N. (2003). Urban development and handing culture down to latecomer: a planning for Beijing Nanchizi preservation area of historical culture (experimental project). 建筑学报*Architectural Journal,* 11, pp. 7-11.

Lin, N. (2004). Insist on the views of practice and development and search for method of protecting historic street areas. 北京规划建设*Beijing Planning Review,* 2.

Lin, Q. (1847/1984). *Hong xue yin yuan tu ji* (drawings by C. Jiang *et al.* 鸿雪因缘图记). China: Beijing Ancient Books Publishing House.

Liu, L.G. (1989). *Chinese architecture*. London: Academy Editions.

Liu, W. (1992). *Users' report on the new courtyard housing in Juer Hutong.* Unpublished report (Chinese edition), Tsinghua University, Beijing, China.

Liu, X. (2002a). The origins of Chinese architecture. In X. Fu *et al., Chinese architecture* (pp. 11-31). New Haven: Yale University Press.

Liu, X. (2002b). The Qin and Han dynasties. In X. Fu *et al., Chinese architecture* (pp. 33-59). New Haven: Yale University Press.

Liu, Z. (1990). *A brief history of Chinese residential architecture: cities, houses, gardens.* Beijing: China Architecture and Building Press.

Lo, K.-Y. (2005). Traditional Chinese architecture and furniture: a cultural interpretation. In R.G. Knapp and K.-Y. Lo (Eds.), *House home family: living and being Chinese* (pp. 161-203). Honolulu: University of Hawai'i Press.

Logan, J.R. (2002). Three challenges for the Chinese city: globalization, migration, and market reform. In J.R. Logan (Ed.), *The new Chinese city: globalization and market reform* (pp. 3-21). Oxford: Blackwell Publishers.

Lu, J. (2004). Rewriting Beijing: a spectacular city in Qiu Huadong's urban fiction. *Journal of Contemporary China*, 13(39), pp. 323-338.

Lu and He (2004). *Beijing Xinjiekou old neighbourhood survey report*. School of Architecture, Tsinghua University, Beijing, China.

Lü, J. (1993). Beijing's old and dilapidated housing renewal (phase 1). *Building in China,* 6(3-4), pp. 24-35.

Luó, Z.W. (1998). Collected works of Luo Zhewen on the preservation of famous historical and cultural city and ancient architecture (罗哲文历史文化名城与古建筑保护文集, Chinese edition). Beijing: China Architectural Industry Publishing House.

Luó, Z.W. (2006). Feng Shui theory and ancient Chinese architectural planning and construction (foreword). In Z.Z. Luò, *Feng shui and modern residences* (风水学与现代家居, Chinese edition, pp. 1-19). Beijing: China City Press.

Luò, Z.Z. (2006). *Feng shui and modern residences* (风水学与现代家居, Chinese edition). Beijing: China City Press.

Lutz, J.G. (1971). *China and the Christian colleges 1850-1950*. Ithaca, NY: Cornell University Press.

Lyle, J.T. (1994). *Regenerative design for sustainable development*. Toronto: John Wiley and Sons.

Ma, B. (1993). *Quadrangles of Beijing* (北京四合院). China: Beijing Arts and Photography Publishing House.

Ma, B. (1999). *The architecture of the quadrangle in Beijing* (北京四合院建筑, Chinese edition). China: Tianjin University Press.

Mann, D. A. (1984). Housing in a state of conflict: tradition and modernization in the People's Republic of China. *Ekistics,* 51(307), pp. 349-354.

Mann, S. (1997). *Precious records: women in China's long 18th century*. Stanford: Stanford University Press.

Marcus, C.C. (2004). Unknown pleasures: many courtyards are permanently locked because they were never intended for use. *Landscaping,* February, pp. 12-16.

Marquand, R. (2001). Why old Beijing's crumbling courtyards face extinction. *Christian Science Monitor,* 93(75), p. 12.

Marshall, G. (Ed.) (1994). *The concise Oxford dictionary of sociology*. Oxford: Oxford University Press.

Marshall, G. (Ed.) (1998). *Oxford dictionary of sociology* (2nd ed.). Oxford: Oxford University Press.

Martin, J. L. and March, L. (1972). *Urban space and structures*. Cambridge, UK: Cambridge University Press.

McDonald, H. (2004) *Beijing invites foreign buyers into old courtyard homes*. Retrieved March 5, 2007 from http://www.smh.com.au/articles/2004/05/28/1085641716523.html?from=storyrhs

Mencius (372-289 BCE). *The works of Mencius* (translated and annotated by J. Legge, 1895/1970). New York: Dover Publications.

Mercer, C. (2006). Local policies for cultural diversity: systems, citizenship, and governance: with an emphasis on the UK and Australia. In UNESCO's report (September 20, 2006), *Local policies for cultural diversity* (pp. 72-85). UNESCO, UCLG and Barcelona City Council.

Merchant, C. (2004). *Reinventing Eden: the fate of nature in Western culture*. New York: Routledge.

Metzger, T.A. (1991). Confucian thought and the modern Chinese quest for moral autonomy. In S. Krieger and R. Trauzettel (Eds.), *Confucianism and the modernization of China* (pp. 266-306). Mainz, Germany: Hase and Koehler Verlag.

Meyer, J.F. (2001). Salvation in the garden: Daoism and ecology. In N.J. Girardot, J. Miller, and X. Liu (Eds.), *Daoism and ecology: ways within a cosmic landscape* (pp. 219-236). Cambridge, MA: Harvard University.

Michael, F. (1977). *Mao and the perpetual revolution*. New York: Barron's Woodbury.

Mitchell, K. (2010). Learning from traces of past living: courtyard housing as precedent and project. In N.O. Rabbat (Ed.), *The courtyard house: from cultural reference to universal relevance* (pp. 223-237). London: Ashgate.

Mohanty, M. (1978). *The political philosophy of Mao Tse-tung*. Delhi, India: Macmillan.

Moore, R.C. (1995). Children gardening: first steps towards a sustainable future. *Children's Environments, 12*(2), pp. 222-232.

Moro, M. and Spirandelli, B. (2011). *The ecological house: sustainable architecture around the world*. Vercelli: White Star.

Mozi (1929/1978). *The ethical and political works of Motse* (translated by W. P. Mei, who omits the Canons, Daqu, Xiaoqu, and the military chapters. English translation of the Canons is based on that in A.C. Graham's *Later Mohist Logic, Ethics, and Science*. Hong Kong: Chinese University Press.

Munro, D. (1995) Sustainability: rhetoric or reality? In T.C. Trzyna and J.K. Osborn (Eds.), *A sustainable world*. CA: International Center for the Environment and Public Policy.

Nan, X. (2003). Nanchizi can never be returned again (再也回不去的南池子). Contributor: Y. Bai. 南方周末 *Southern Weekend*, October 27. Retrieved January 7, 2009 from: http://www.southcn.com/travel/custom/xsj/200310270617.htm

Nanchizi Project Construction Headquarters (2004). Brief on the Nanchizi rebuilding situation of culture protection areas (南池子文保区修缮改建前后情况介绍). 北京规划建设*Beijing City Planning and Construction Review*, 2, pp. 98-100.

Nee, V. (1996). The emergence of a market society: changing mechanisms of stratification in China. *American Journal of Sociology*, 101, pp. 908-949.

Neville, R.C. (1998). Orientation, self, and ecological posture. In M.E. Tucker and J. Berthrong (Eds.), *Confucianism and ecology: the interrelation of heaven, earth, and humans* (pp. 265-271). Cambridge, MA: Harvard University Press.

Nilsson, J. (1998). *Problems and possibilities in today's urban renewal in the old City of Beijing*. Department of Urban Planning and Design, Tsinghua University, Beijing, China.

Noble, A.G. (2003). Patterns and relationships of Indian houses. In R.G. Knapp (Ed.), *Asia's old dwellings: tradition, resilience, change* (pp. 39-69). New York: Oxford University Press.

Nurse, K. (2006). *Culture as the fourth pillar of sustainable development*. Paper prepared for Commonwealth Secretariat, Malborough House, Pall Mall, London, UK. Retrieved September 27, 2011 from: http://www.fao.org/SARD/common/ecg/2785/en/Cultureas4thPillarSD.pdf

OECD (2001). *Sustainable development: critical issues*. Paris: Organization for Economic Cooperation and Development.

Oliver, P. (Ed.). (1997a). *Encyclopedia of vernacular architecture of the world: Theories and principles (Vol. 1)*. Cambridge: Cambridge University Press.

Oliver, P. (1997b). Recording and documentation. In P. Oliver (Ed.), *Encyclopedia of vernacular architecture of the world: theories and principles* (Vol. 1) (pp. 56-58). Cambridge: Cambridge University Press.

Onions, C.T. (Ed.) (1964). *The shorter Oxford English dictionary*. Oxford: Clarendon Press.

Oppenheim, A. N. (1992). *Questionnaire design, interviewing and attitude measurement*. New York, NY: Continuum.

Orasia (2002). *Suzhou*. Retrieved February 14, 2008 from http://www.asia-planet.net/china/suzhou.htm

Ornelas, M. (2006). Beijing's rapid urbanization and the hutong. *La Voz de Esperanza*, 19(8), pp. 7-9.

Orr, D.W. (2002). *The nature of design: ecology, culture, and human intention*. New York: Oxford University Press.

O'Shea, M. (2011). Arts engagement with sustainable communities: informing new governance styles for sustainable futures. *Culture and Local Governance / Culture et gouvernance locale*, 3(1-2), pp. 29-41.

Outhwaite, W., and Bottomore, T. (Eds.) (1993). *The Blackwell dictionary of twentieth-century social thought*. Oxford: Blackwell.

Oxford Brookes University (2007). *Module handbook for P37639: statistical research using SPSS*. School of Built Environment, Oxford Brookes University, UK.

Özcan, Z., Gültekin, N. T., and Dündar, Ö. (1998). *Sustainability versus development: Mudanya's war of survival as a liveable city*. 38th Congress of the European Regional Science Association, 28 August - 1 September 1998, Vienna, Austria.

Özkan, S. (2006). Preface. In B. Edwards *et al.* (Eds.), *Courtyard housing: past, present, and future* (pp. xvii-xix). New York: Taylor and Francis.

Paehlke, R. (2004). Green at home. *Alternatives Journal*, 30(5), p. 3.

Pallasmaa, J. (2001). Our image culture and its misguided ideas about freedom. *Architectural Record*, 189(1), pp. 51-52.

Pan, G. (2002). The Yuan and Ming dynasties. In X. Fu *et al.*, *Chinese architecture* (pp. 199-259). New Haven: Yale University Press.

Pavlides, E. (1997). Architectural. In P. Oliver (Ed.), *Encyclopedia of vernacular architecture of the world: theories and principles* (Vol. 1) (pp. 12-15). Cambridge: Cambridge University Press.

Pfeifer, G. and Brauneck, P. (2008). *Courtyard houses: a housing typology*. Basel/Boston/Berlin: Birkhäuser Verlag.

Pheng, L.S. (2001). Construction of dwellings and structures in ancient China. *Structural Survey*, 19(5), pp. 262-274.

Pollard, P.H. (1970). *Medium-density courtyard housing in Sydney.* M.T.C.P. thesis, University of Sydney, Australia.

Polyzoides, S., Sherwood, R., and Tice, J. (1992). *Courtyard housing in Los Angeles: a typological analysis* (2nd ed.). New York: Princeton Architectural Press.

Pounds, N.J.G. (1997). Historical. In P. Oliver (Ed.), *Encyclopedia of vernacular architecture of the world: theories and principles* (Vol. 1) (pp. 46-48). Cambridge: Cambridge University Press.

Qi, Z. and Yamashita, T. (2004). Lilong housing renewal in the luwan district of Shanghai and changes of residential environments: a study of lilong housing renewal in Shanghai's old city part. *Journal of Architecture and Planning, 579,* pp. 7-14.

Qiao, Y. (2002). Introduction. In X. Fu *et al., Chinese architecture* (pp. 5-9). New Haven: Yale University Press.

Rabbat, N.O. (Ed.) (2010). *The courtyard house: from cultural reference to universal relevance.* London: Ashgate.

Rajapaksha, I., Nagai, H. and Okumiya, M. (2002). Indoor thermal modification of a ventilated courtyard house in the tropics. *Journal of Asian Architecture and Building Engineering,* 1(1), pp. 87-94.

Randhawa, T. S. (1999). *Indian courtyard house.* New Delhi: Prakash Books.

Raphals, L. (1998). *Sharing the light: representations of women and virtue in early China.* Albany: State University of New York Press.

Rapoport, A. (1969). *House form and culture.* Englewood Cliffs, NJ: Prentice-Hall.

Rapoport, A. (1983). Environmental quality: metropolitan areas and traditional settlements. *Habitat International,* 7(3/4), pp. 37-63.

Rapoport, A. (1997). Behavioural. In P. Oliver (Ed.), *Encyclopedia of vernacular architecture of the world: theories and principles* (Vol. 1) (pp. 16-18). Cambridge: Cambridge University Press.

Rapoport, A. (2005). *Culture, architecture, and design.* Chicago, IL: Locke Science.

Reid, T.R. (2000). *Confucius lives next door: what living in the East teaches us about living in the West.* New York: Vintage Books.

Residential Design and Construction Specifications (1987, 2003, and 2006) (住宅建筑设计规范). Retrieved August 2008 from: http://www.law110.com/lawserve/guihua/1800002.htm   http://www.lzgh.cn/low/19/2006/06/22/1631,19.aspx http://sjzbbs.soufun.com/1310118903~-1~435/40003883_40003883.htm

Reynolds, J.S. (2002). *Courtyards: aesthetic, social, and thermal delight.* New York: John Wiley and Sons.

Reuber, P. (1998). Beijing's hutongs and siheyuan. *Canadian Architect,* 43(10), pp. 40-43.

RIA Novosti (2011). *China becomes world's largest car market.* Retrieved March 10, 2011 from: http://en.rian.ru/world/20090206/120007709.html

Ro, Y.-C. (1998). Ecological implications of Yi Yulgok's cosmology. In M.E. Tucker and J. Berthrong (Eds.), *Confucianism and ecology: the interrelation of heaven, earth, and humans* (pp. 169-186). Cambridge, MA: Harvard University Press.

Roaf, S., Fuentes, M., and Thomas, S. (2007). *Ecohouse* (3rd ed.). Oxford: Architectural Press.

Sassi, P. (2006). *Strategies for sustainable architecture*. Oxon: Taylor and Francis.

Savova, N. (2011). Arm's length and 'hand-shake' policies: community arts alternatives to outcome-based development (insights from Brazil, Bulgaria, and South Africa). *Culture and Local Governance / Culture et gouvernance locale,* 3(1-2), pp. 43-58.

Schefold, R. (1997). Anthropological. In P. Oliver (Ed.), *Encyclopedia of vernacular architecture of the world: theories and principles* (Vol. 1) (pp. 6-8). Cambridge: Cambridge University Press.

Schell, O. (1995). China: the end of an era. *The Nation,* July 17/24, pp. 84-98.

Schipper, K. (2001). Daoist ecology: the inner transformation. A study of the precepts of the early Daoist ecclesia. In N.J. Girardot, J. Miller, and X. Liu (Eds.), *Daoism and ecology: ways within a cosmic landscape* (pp. 79-93). Cambridge, MA: Harvard University Press.

Schinz, A. (1989). *Cities in China*. Stuttgart: Borntraeger Science Publishers.

Schmid, P. (1987). Architects: let us make the world a better place! *Open House International,* 12(1), pp. 48-54.

Schram, S.R. (Trans.) (1967). *Quotations from chairman Mao Tse-tung* (2nd ed.). Peking: Foreign Languages Press.

Schram, S.R. (1971). *Mao Tse-tung*. Baltimore, MD: Penguin Books.

Schrift, M. (2001). *Biography of a chairman Mao badge*. New Jersey: Rutgers University Press.

Schwartz, B.I. (1985). *The world of thought in ancient China*. Cambridge, MA: The Belknap Press of Harvard University Press.

Scruton, R. (1996). *A dictionary of political thought* (2nd ed.). London: Macmillan.

Shang, K. and Yang, L. Y. (1982). Traditional courtyard houses and low-rise high density (传统庭院式住宅与低层高密度). 建筑学报*Architectural Journal,* 5, pp. 51-60.

Shao, Z. (2005). *The art of Suzhou classical gardens* (苏州古典园林艺术, photography by R. Zhou and S. Gao). Beijing: China Forestry Publishing House.

Sheng, M. (2008). *Report on the property management of housing estates in the eastern district of Beijing* (关于东城区居住小区物业管理情况调研报告). Retrieved January 28, 2009 from: http://renda.bjdch.gov.cn/n4279063/n4279168/n4279801/4656244.html

Shu, Y. (2004). The gain and loss of Nanchizi (南池子的得与失). 北京规划建设 *Beijing City Planning and Construction Review,* 2, pp. 114-116.

Simons, D. and Dang, S.R. (2006). *International perspectives on cultural indicators: a review and compilation of cultural indicators used in selected projects*. Centre of Expertise on Culture and Communities Creative City Network of Canada/Simon Fraser University.

Sinha, A. (1994). The centre as void: courtyard dwellings in India. *Open House International,* 19(4), pp. 28-35.

Sit, V.F.S. (1999). Social areas in Beijing. *Geografiska Annaler,* 81B(4), pp. 203-221.

Smith, P. (2008) China's quake: why did so many schools collapse? Earthquake experts say the collapsed schools may be a sign of poor construction despite adequate building codes. *Christian Science Monitor*, May 14, 2008. Retrieved January 25, 2012 from: http://www.csmonitor.com/World/Asia-Pacific/2008/0514/p06s05-woap.html

Smith, R.J. (1994). *China's cultural heritage: the Qing dynasty, 1644-1912* [sic]. Boulder: Westview Press.

Song, X.G. (1988). *A discussion of the conservation and reconstruction of the courtyard housing residential area in the old city of Beijing*. Master's thesis, Tsinghua University, Beijing, China.

Spalding, D. (n.d.). *Ghosts among the ruins: urban transformation in contemporary Chinese art*. Retrieved March 5, 2007 from http://www.cca.edu/sightlines/pdf/dspalding.pdf

Spaling, H. and Dekker, A. (1996). *Cultural sustainable development: concepts and principles*. Retrieved September 20, 2009 from: http://www.asa3.org/ASA/PSCF/1996/PSCF12-96Spalling.html

Sprenger, A. (1991). Confucius and modernization in China: an educational perspective. In S. Krieger and R. Trauzettel (Eds.), *Confucianism and the modernization of China* (pp. 454-472). Mainz, Germany: Hase and Koehler Verlag.

Stanborough, M. (2011). The link between: culture and sustainability in municipal planning. *Culture and Local Governance / Culture et gouvernance locale*, 3(1-2), pp. 95-100.

Stang, A. and Hawthorne, C. (2010). *The green house: new directions in sustainable architecture*. NY: Princeton Architectural Press.

State Council (1990). Provisional ordinance for urban state owned land use right transfer. In Land Policy Section of Beijing Municipal Real Estate Management Bureau (Eds.) (1993), *Land management policy and legislation in Beijing* (pp. 60-66). Beijing: Land Policy Section of Beijing Municipal Real Estate Management Bureau.

Steinhardt, N.S. (2002). Introduction. In X. Fu *et al.*, *Chinese architecture* (pp. 1-4). New Haven: Yale University Press.

Stepanchuk, C. and Wong, C. (1991). *Mooncakes and hungry ghosts: festivals of China*. CA: China Books and Periodicals.

Stewart, W.F.R. (1970). *Children in flats: a family study*. National Society for the Prevention of Cruelty to Children (NSPCC).

Sun, D. (2002). The Qin dynasty. In X. Fu *et al.*, *Chinese architecture* (pp. 261-343). New Haven: Yale University Press.

Sun, F. (2007). *Baobao zhuazhou, forecasting the future and nature of the child* (宝宝抓周 五行测志趣). Retrieved June 24, 2010 from: http://baobao.sohu.com/20070914/n252148828.shtml

Sun, H. and Gong, Q. (1989). Initial experiment of detailed planning in old city regeneration schemes: Suzhou Tongfang Xiang regeneration planning (旧城改造详细规划中的土地区划初探：苏州桐芳巷改造规划). 城市规划*City Planning Review*, pp. 10-15.

Sustainable Development Research Institute (1998). *Social capital formation and institutions for sustainability*. Workshop proceedings (November 16 - 17, 1998) prepared by Asoka Mendis. Vancouver: Sustainable Development Research Institute. Retrieved November 6, 2011 from http://www.williambowles.info/mimo/refs/soc_cap.html

Suzhou Housing Management Bureau (2004). *Ancient residences in Suzhou* (苏州古民居). Shanghai, China: Tongji University Press.

Suzhou Old City Construction Office (1991). *Number 20: regarding the application for establishing Tongfangxiang housing regeneration experiment* (关于桐芳巷住宅小区改造试点申请立项的报告). Suzhou Urban and Rural Construction Archives.

Suzhou Sujing Real Estate Development Corporation (1996). *Regarding the application for establishing the location of Shizilin/Shilinyuan housing regeneration project* (关于申办狮子林/狮林苑小康住宅小区改造项目规划定点的报告). Suzhou Urban and Rural Construction Archives.

Suzhou Tongfangyuan Housing Estate Redevelopment Office (1992). *Several issues regarding the technical design of Tongfangxiang's northeast patch* (关于桐芳巷东北片技术设计工作方面应掌握的几个问题). Suzhou Urban and Rural Construction Archives.

Suzhou Tongfangyuan Housing Estate Redevelopment Office (1994). *Minutes of the meeting regarding Suzhou Tongfangxiang housing experiment in its southwest and northwest patches* (苏州市桐芳巷住宅试点小区西南，西北片规划调整论证会纪要). Suzhou Urban and Rural Construction Archives.

Suzhou Tongfangyuan Housing Estate Redevelopment Office (1995). *Minutes of the meeting regarding Tongfangxiang housing estate's landscaping design in its southwest and northwest patches* (桐芳巷住宅小区西南，西北片绿化方案论证会纪要). Suzhou Urban and Rural Construction Archives.

Suzhou Urban Planning Bureau (1997). *Number 242: submission of the validation of Shilinyuan housing estate design proposal* (狮林苑设计方案审定意见书). Suzhou Urban and Rural Construction Archives.

Suzhou Urban Planning Committee (1992). *Number 27: regarding the approval of Tongfangxiang housing estate regeneration project* (关于桐芳巷住宅小区改造项目任务书的批复). Suzhou Urban and Rural Construction Archives.

Suzhou Urban Planning Committee (1994). *Number 452: regarding the approval of the adjustment of Tongfangxiang housing estate regeneration project* (关于同意调整桐芳巷住宅小区改造项目计划的批复). Suzhou Urban and Rural Construction Archives.

Tan, Y. (1994). Social aspects of Beijing's old and dilapidated housing renewal. *China City Planning Review* (English edition), 10(4), pp. 45-55.

Tan, Y. (1997). *Research into reconstruction of the old residential area of Beijing: from the residents' perspective.* PhD thesis, Tsinghua University, Beijing, China.

Tan, Y. (1998). *Relocation and the people: a research on neighbourhood renewal in the old city of Beijing.* Department of Urban Planning and Design, Tsinghua University School of Architecture, Beijing, China.

Terrill, R. (1999). *Mao: a biography.* Stanford, CA: Stanford University Press.

Theis, J.S. (2005). Milton's principles of architecture. *English Literary Renaissance,* 35(1), pp. 102-122.

Throsby, D. (2008). *Culture in sustainable development: insights for the future implementation of article 13.* Paper for the UNESCO Convention on the Protection and Promotion of the Diversity of Cultural Expressions.

Together Foundation and UNCHS (2002). *The protection and upgrading of the ancient city of Suzhou, China.* Retrieved December 18, 2006 from: http://www.ucl.ac.uk/dpu-projects/drivers_urb_change/urb_society/pdf_cult_ident/HABITAT_BestPractice_Protection_Suzhou_China.pdf

Trapp, M. (2002?). *A Comparison of the relationship between urban development and the natural environment in Beijing, Tokyo, and Curitiba.* Retrieved March 5, 2007 from http://www.rd.msu.edu/ugrad/greenpieces/2003/Trapp.pdf

Travel China and the World (2010-11-01). *Housing, apartments, houses/villas, construction quality in China/Beijing.* Retrieved July 2, 2011 from: http://www.lonelyplanet.com/travelblogs/718/87431/Housing,+Apartments,+Houses-Villas,+Construction+quality+in+China+-+Beijing?destId=355904

Tsinghua University (1994). *Beijing Ju'er Hutong new courtyard housing experiment.* Beijing, China: Institute of Architectural and Urban Studies, Tsinghua University.

Tu, W. (1998a). Beyond the enlightenment mentality. In M.E. Tucker and J. Berthrong (Eds.), *Confucianism and ecology: the interrelation of heaven, earth, and humans* (pp. 3-21). Cambridge, MA: Harvard University Press.

Tu, W. (1998b). The continuity of being: Chinese visions of nature. In M.E. Tucker and J. Berthrong (Eds.), *Confucianism and ecology: the interrelation of heaven, earth, and humans* (pp. 105-121). Cambridge, MA: Harvard University Press.

Tucker, M.E. (1998). The philosophy of ch'i as an ecological cosmology. In M.E. Tucker and J. Berthrong (Eds.), *Confucianism and ecology: the interrelation of heaven, earth, and humans* (pp. 187-207). Cambridge, MA: Harvard University Press.

Tucker, M.E. and Berthrong, J. (Eds.) (1998). *Confucianism and ecology: the interrelation of heaven, earth, and humans.* Cambridge, MA: Harvard University Press.

Tung, A.M. (2003). Erasing old Beijing: a conservation tragedy. *World Monuments Icon,* pp. 38-45.

Tylor, E.B. (1871/1920). *Primitive culture.* New York: J.P. Putnam's Sons.

Uhalley, S. Jr. (1975). *Mao Tse-tung: a critical biography*. New York: New Viewpoints.

Ujam, F. (2006). The cosmological genesis of the courtyard house. In B. Edwards *et al.* (Eds.), *Courtyard housing: past, present, and future*. New York: Taylor and Francis.

Ulrich, R.S. (1984). View through a window may influence recovery from surgery. *Science, 224*, pp. 420-421.

UN (2002?). *Demographic and social trends affecting families in the south and central Asian region*. Retrieved May 30, 2011 from: http://www.un.org/esa/ socdev/family/Publications/mtdesilva.pdf

UN World Commission on Environment and Development (1987). *Our common future: the Brundtland report*. Oxford: Oxford University Press.

UN World Commission on Culture and Development (1995). *Our creative diversity*. Paris: UNESCO.

UNCED (1992). Rio Declaration. United Nations Conference on Environment and Development, Rio de Janeiro, June 3-14, 1992.

UNESCO (2002). *Universal declaration on cultural diversity*. Paris: UNESCO.

UNESCO (2006). *Local policies for cultural diversity*. Paris: UNESCO.

UNESCO (2009). *Convention for the safeguarding of the intangible cultural heritage*. Paris: UNESCO.

UNESCO-UIS (2009). *2009 UNESCO framework for cultural statistics*. Montreal: UNESCO Institute for Statistics.

UNESCO World Bank (2000). *China-cultural heritage management and urban development: challenge and opportunity*. International Conference on Cultural Heritage Management and Urban Development: Challenge and Opportunity, July 5-7, 2000, Beijing, China.

UNESCO World Report (2009). *Investing in cultural diversity and intercultural dialogue*. Paris: UNESCO.

United Cities and Local Governments (2006a). *Agenda 21 for culture: working group on culture - activities (Circular 1)*.

United Cities and Local Governments (2006b). *Agenda 21 for culture: working group on culture (Circular 2)*.

United Cities and Local Governments (2006c). *Key ideas on the Agenda 21 for culture*.

United Cities and Local Governments (2006d). *Advice on local implementation of the Agenda 21 for culture*.

United Cities and Local Governments (2009). *Agenda 21 for culture: institutional innovation in local policies for cultural diversity and intercultural dialogue*.

United Cities and Local Governments (2010). *Culture: fourth pillar of sustainable development*. United Cities and Local Governments (UCLG) Policy Statement, November 17, 2010.

United Nations General Assembly (2005). *2005 World Summit outcome, resolution A/60/1*. Adopted by the General Assembly on 15 September 2005.

Ursic, M. (2011). The diminished importance of cultural sustainability in spatial planning: the case of Slovenia. *Culture and Local Governance / Culture et gouvernance locale*, 3(1-2), pp. 101-116.

Van Elzen, S. (2010). *Dragon and rose garden: art and power in China*. Hong Kong: Timezone 8/Modern Chinese Art Foundation.

Wagner, J.D. (2008). *Green remodeling: your start toward an eco-friendly home* (The Green House). NJ: Creative Homeowner.

Walters, D. (1991). *The Feng Shui handbook: a practical guide to Chinese geomancy and environmental harmony*. San Francisco, CA: Thorsons.

Wang, B. (1997). *Yellow emperor's canon of internal medicine* (黄帝内经 translated by L. Wu and Q. Wu). Beijing: China Science and Technology Press.

Wang, J. (2003). *Beijing record* (城记Chinese edition). China: SDX Joint Publishing.

Wang, J. (2006). *History and reality of old Beijing: hutong and Beijing* (北京旧城的历史与现实：胡同与北京). Speech made at Beijing Cultural Heritage Protection Center. Retrieved February 13, 2008 from: http://www.bjchp.org/oldbeijing/3/a_wangjun.asp

Wang, J. (2007). Old city of Beijing will not demolish hutong and siheyuan in large-scale （北京旧城区将不再成片推倒胡同四合院了）. 瞭望新闻周刊 *Observers News Weekly*, December 9, 2007. Retrieved February 13, 2008 from: http://news.xinhuanet.com/mrdx/2007-12/09/content_7220797.htm

Wang, J. (2011). *Beijing record: a physical and political history of planning modern Beijing* (English edition). Singapore/USA/UK/India/China: World Scientific Publishing.

Wang, J.C. (2005). House and garden: sanctuary for the body and the mind. In R.G. Knapp and K.-Y. Lo (Eds.), *House home family: living and being Chinese* (pp. 73-97). Honolulu: University of Hawai'i Press.

Wang, Q. (1999). *Beijing siheyuan* (北京四合院, Chinese edition). Beijing: China Bookstore.

Wang, S. (2004). Discussion on protecting pilot project of Nanchizi (保护历史文化街区的价值取向原则—兼议南池子保护试点工程). 北京规划建设 *Beijing City Planning and Construction Review*, 2, pp. 105-108.

Wang, W. (2006). *Protecting Beijing's cultural symbol: siheyuan* (保护北京文化的表征-四合院). 光明日报 Guangming Daily, February 20, 2006.

Wang, Y.C. (2003). Measures taken for the preservation of Suzhou historical street blocks (从" 桐芳巷" 到" 新天地" — 谈苏州历史街区保护对策). 规划师 Planners, 19(6), pp. 20-23.

Wang, Y.P. (1995). Public sector housing in urban China 1949-1988: the case of Xian. *Housing Studies*, 10(1), pp. 57-82.

Wang, Y.P. and Murie, A. (1999). Commercial housing development in urban China. *Urban Studies*, 36(9), pp. 1475-1494.

Watson, S. and Studdert, D. (2006). *Markets as sites for social interaction: spaces of diversity*. London: UK. Joseph Rowntree Foundation Report no. 1940.

Watts, J. (2007). Property boom threatens old Beijing. *The Guardian,* April 17, p. 17.

Weggel, O. (1991). Between Marxism and meta-Confucianism: China on her way "back to normality." In S. Krieger and R. Trauzettel (Eds.), *Confucianism and the modernization of China* (pp. 400-414). Mainz, Germany: Hase and Koehler Verlag.

Weller, R.P. and Bol, P.K. (1998). From heaven-and-earth to nature: Chinese concepts of the environment and their influence on policy implementation. In M.E. Tucker and J. Berthrong (Eds.), *Confucianism and ecology: the interrelation of heaven, earth, and humans* (pp. 313-341). Cambridge, MA: Harvard University Press.

Wheelwright, P. (2000). *The problem of sustainability: a response* (Response to Molecules, Money, and Design by Mark Jarzombek). *Thresholds 20, MIT Journal,* Fall 2000. Retrieved August 4, 2004 from: http://www.pmwarchitects.com/ac_tech.htm

Whitehand, J.W.R. and Gu, K. (2006). Research on Chinese urban form: retrospect and prospect. *Progress in Human Geography*, 30(3), pp. 337-355.

Williams, D.E. (2007). *Sustainable design: ecology, architecture, and planning.* NJ: Wiley.

Wilson, D. (1979). *Mao: the people's emperor.* London: Hutchinson.

Wood, F. (1987). Traditional domestic architecture and its modern fate. *Open House International,* 12(1), pp. 42-47.

World Bank (1992). *China: implementation options for urban housing reform* (A World Bank Country Study). Washington, DC: The World Bank.

Wu, F. (2004). Transplanting cityscapes: the use of imagined globalization in housing commodification in Beijing. *Area*, 36(3), pp. 227-234.

Wu, F. (2005). Rediscovering the 'gate' under market transition: from work-unit compounds to commodity housing enclaves. *Housing Studies*, 20(2), pp. 235-254.

Wu, L. (1986). *A brief history of ancient Chinese city planning.* In Urbs et Regio. Kassel: Gesamthochschulbibliothek.

Wu, L. (1989). Rehabilitation of residential areas in the old city of Beijing. *Architectural Journal,* July, pp. 11-18 (Chinese edition).

Wu, L. (1991a). Rehabilitation of residential areas in the old city of Beijing. *Architectural Journal,* February, pp. 7-13 (Chinese edition).

Wu, L. (1991b). Rehabilitation of residential areas in the old city of Beijing. *Architectural Journal.* December, pp. 2-12 (Chinese edition).

Wu, L. (1991c). Rehabilitation in Beijing. *Habitat International,* 15(3), pp. 51-66.

Wu, L. (1991d). Towards a theory of regional reality. *Habitat International,* 15(3), pp. 3-9.

Wu, L. (1994). The future direction of the experiments. *Beijing Ju'er Hutong New Courtyard Housing Experiment.* Beijing, China: Institute of Architectural and Urban Studies, Tsinghua University.

Wu, L. (1999). *Rehabilitating the old city of Beijing: a project in the Ju'er Hutong neighbourhood*. Vancouver, BC: University of British Columbia Press.

Wu, L. (2005a). Special discussion on Beijing City Master Plan 2004-2020: research on Beijing's old city protection (part 1) (北京城市总体规划修编 2004-2020年专题:北京旧城保护研究(上篇)). 北京规划建设*Beijing City Planning and Construction Review*, 1, pp. 20-28.

Wu, L. (2005b). Special discussion on Beijing City Master Plan 2004-2020: research on Beijing's old city protection (part 2) (北京城市总体规划修编 2004-2020年专题:北京旧城保护研究(下篇)). 北京规划建设 *Beijing City Planning and Construction Review*, 2, pp. 65-72.

Wu, N.I. (1968). *Chinese and Indian architecture: the city of man, the mountain of God, and the realm of the immortals*. London: Studio Vista.

Wu, Y. (1983). *Treasury of Wu Youru's drawings* (吴有如画宝). China: Shanghai Bookstore Publisher.

Xinhua News Agency (2010-07-13). *Culture and Edu: 316 Confucius Institutes established worldwide*. Retrieved May 8, 2011 from: http://news.xinhuanet. com/english2010/culture/2010-07/13/c_13398209.htm

Xu, P. (1998). Feng-Shui models structured traditional Beijing courtyard houses. *Journal of Architectural and Planning Research*, 15(4), pp. 271-282.

Xu, Y. (2000). *The Chinese city in space and time: the development of urban form in Suzhou*. Honolulu: University of Hawai'i Press.

Yan, X.W. and Marans, R.W. (1995). Perceptions of housing in Beijing. *Third World Planning Review,* 17(1), pp. 19-37.

Yan, Y. (2005). Making room for intimacy: domestic space and conjugal privacy in rural north China. In R.G. Knapp and K.-Y. Lo (Eds.), *House home family: living and being Chinese* (pp. 373-395). Honolulu: University of Hawai'i Press.

Yang, L. (2008). *Yang Li talks about Yi Jing* (杨力讲易经Chinese edition). China: Beijing Science and Technology Publishing House.

Yeang, K. (1995). *Designing with nature: the ecological basis for architectural design*. Columbus, OH: Mcgraw-Hill.

Yoon, S. and Choi, K. (2001). *Comparative study on Korean and Chinese housing based on cultural patterns: from the 15th century to the present*. Graduate School of Techno Design, Kookmin University, Department of Interior Design, Seoul, Korea.

Yu, D. (1991). The concept of "great harmony" in the Book of Changes (Zhou Yi): Confucian philosophical theories on conflict and harmony. In S. Krieger and R. Trauzettel (Eds.), *Confucianism and the modernization of China* (pp. 51-62). Mainz, Germany: Hase and Koehler Verlag.

Yu, S. (2007). *Suzhou's old city protection and its historic and cultural values* (苏州古城保护及其历史文化价值 Chinese edition). China: Shanxi People's Education.

Yuan, X. and Gong, J. (2004). *The classical gardens of Suzhou: world cultural heritage* (translated by H. Wang and S. Hua). Nanjing: Jiangsu People's Publishing House.

Yuan, Y. (2005). *Preserving the soul of Beijing.* Retrieved March 5, 2007 from http://www.chinatoday.com.cn/English/e2005/e200502/p28.htm

Yue, A., Khan, R., and Brook, S. (2011). Developing a local cultural indicator framework in Australia: a case study of the city of Whittlesea. *Culture and Local Governance / Culture et gouvernance locale,* 3(1-2), pp. 133-149.

Yung, P. and Yip, B. (2010). Construction quality in China during transition: a review of literature and empirical examination. *International Journal of Project Management,* 28(1), pp. 79-91.

Zhang, D. (1994). *New courtyard houses of Beijing: direction of future housing development.* MA dissertation in Urban Design, Joint Centre for Urban Design, School of Architecture, Oxford Brookes University, UK.

Zhang, D. (2006). New courtyard houses of Beijing: direction of future housing development. *Urban Design International,* 11(3/4), pp. 133-150.

Zhang, D. (2009). *Schoolyard gardening as multinaturalism: theory, practice, and product.* Saarbrücken, Germany: VDM Verlag.

Zhang, D. (2009/2010/2011). *Courtyard houses of Beijing: past, present, and future.* Saarbrücken, Germany: VDM Verlag.

Zhang, J. (2001). A declaration of the Chinese Daoist Association on global ecology (translated by David Yu). In N.J. Girardot, J. Miller, and X. Liu (Eds.), *Daoism and ecology: ways within a cosmic landscape* (pp. 361-372). Cambridge, MA: Harvard University Press.

Zhang, L. (Ed.) (1993). *Building in China,* 6(3-4). Beijing: China Building Technology Development Center.

Zhang, L. (2010). *In search of paradise: middle-class living in a Chinese metropolis.* Ithaca, NY: Cornell University Press.

Zhang, W. (2005). Investigation on the urban elderly's perception of residential environment (in Shanghai). *Psychological Science,* 28(155), pp. 681-682.

Zhang, X.Q. (1997). Chinese housing policy 1949-1978: the development of a welfare system. *Planning Perspectives,* 12, pp. 433-455.

Zhang, Y., and Fang, K. (2003). Politics of housing redevelopment in China: the rise and fall of the Ju'er Hutong project in inner-city Beijing. *Journal of Housing and the Built Environment,* 18, pp. 75-87.

Zhao, Y. (2000). *Nirvana of tradition: the rebirth of courtyard housing in Beijing.* MPhil thesis, Department of Architecture, Planning and Allied Arts, Faculty of Architecture, University of Sydney, Australia.

Zheng, J. (2005). Rescue hutong (Chinese edition). *Ming Bao* (Brightness Newspaper) *Saturday Supplement,* 504, pp. 21-30.

Zheng, L. (1995). *Urban renewal in Beijing: observation and analysis.* Master of Architecture thesis, School of Architecture, McGill University, Canada.

Zhou, B. (2006). *Beijing rental survey: private housing sold by non-property owners* (北京经租房调查:被非产权人出售的私房). 法制早报 Legality Morning Newspaper, October 23, 2006.

Zhou, M. and Logan, J.R. (2002). Market transition and the commodification of housing in urban China. In J.R. Logan (Ed.), *The new Chinese city: globalization and market reform* (pp. 137-152). Oxford: Blackwell Publishers.

Zhou, X. (1998). Suzhou: the oriental Venice. *UNESCO,* 101, pp. 13-14.

Zhu, J. (2009). *Architecture of modern China: a historic critique.* London and New York: Routledge.

Zhu, J. and Fu, Z. (1988). Planning and design concept for the renewal of number 265, Deshengmen Nei Dajie (德内大街265号更新规划设计构思). 北京规划建设Beijing City Planning and Construction Review, 3, pp. 18-21.

Zhu, X.D, Huang L., and Zhang, X. (2000). *Housing and economic development in Suzhou, China: a new approach to deal with the inseparable issues.* Seminal paper from Joint Center for Housing Studies, Harvard University.

Zhuangzi (c.369-286 BCE). *The complete works of Chuang Tzu* (translated by B. Watson, 1968). New York: Columbia University Press.

Zhuangzi (c.369-286 BCE) and the Central Harmony. *The second book of the Tao* (compiled and adapted by S. Mitchell, 2009). New York: Penguin Press.

Zou, M.Y. (2002). *Feng Shui* (风生水起: 陽宅天數富貴興旺法, Chinese edition). Taipei, Taiwan: Kaixin Publishing.

Index